How to Solve Organic Reaction Mechanisms

How to Solve Organic Reaction Mechanisms

A Stepwise Approach

MARK G. MOLONEY

Fellow and Tutor in Chemistry at St Peter's College
and Professor of Chemistry,
University of Oxford, UK

This edition first published 2015
© 2015 John Wiley & Sons, Ltd

Registered Office
John Wiley & Sons, Ltd, The Atrium, Southern Gate, Chichester, West Sussex, PO19 8SQ, United Kingdom

For details of our global editorial offices, for customer services and for information about how to apply for permission to reuse the copyright material in this book please see our website at www.wiley.com.

Library of Congress Cataloging-in-Publication Data

Moloney, Mark G.
How to solve organic reaction mechanisms : a stepwise approach / Mark G. Moloney.
     pages   cm
Originally published as: Reaction mechanisms at a glance (Malden, Mass. : Blackwell Science, 2000).
Includes index.
ISBN 978-1-118-40159-0 (pbk.)
1. Organic reaction mechanisms.   I. Title.
   QD502.5.M65 2015
   547′.2–dc23
                                        2014024070

A catalogue record for this book is available from the British Library.

ISBN: 9781118401590

Set in 10/12pt Helvetica Condensed by SPi Publisher Services, Pondicherry, India

Printed in Singapore by C.O.S. Printers Pte Ltd

1   2015

Contents

Preface

This book is an upgraded version of *Reaction Mechanisms at a Glance*, first published in 2000. That book was an attempt to demonstrate that there is indeed an underlying set of rules suitable to working out plausible reaction mechanisms in organic chemistry and which can be grasped with a little effort. More importantly, the use of these rules in a systematic fashion substantially reduces the burden on memory! This version has an expanded set of fully worked problems and a new chapter which applies the problem-solving strategy to ligand-coupling reactions using transition metals. The latter is an addition which represents the exceptional growth and importance of this chemistry, and its widespread application in diverse areas of chemical science.

I would like to dedicate this book to my wife Julie and all the members of my family.

Mark Moloney
2014

Abbreviations

Ac	Acetyl ($CH_3C(O)-$)
cat.	catalytic
Δ	Heat
DMF	*N,N*-Dimethylformamide (Me_2NCHO)
MCPBA	meta-chloroperbenzoic acid
PPA	Polyphosphoric acid
THF	Tetrahydrofuran
p-TsCl	*p*-Toluenesulfonyl chloride (*p*-$MeC_6H_4SO_2Cl$)
p-TsOH	*p*-Toluenesulfonic acid (*p*-$MeC_6H_4SO_3H$)
py	Pyridine (C_5H_5N)
EWG	Electron withdrawing group
dil.	dilute
conc.	concentrated

About the companion website

This book is accompanied by a companion website:

www.wiley.com/go/moloney/mechanisms

The website includes worked supplementary questions.

Introduction

There are many organic chemistry texts, old and new, which cover material from the fundamental to the advanced. Most texts, of course, are factually based, but students seem to find considerable difficulty in the application of this factual knowledge to the solution of problems, and all too often attempt to rely on memory alone. However, the sheer volume of material to be committed to memory presents a considerable burden, and the temptation to give up on the subject almost at the outset can be very strong. This book attempts to demonstrate that a general problem-solving strategy is indeed applicable to many of the common reactions of organic chemistry. It develops a checklist approach to problem-solving, using mechanistic organic chemistry as its basis, which is applicable in a wide variety of situations. It aims to show that logical and stepwise reasoning, in combination with a good understanding of the fundamentals, is a powerful tool to apply to the solution of problems.

Philosophy of the book

This book is not a 'fill in the box' text, nor does it have detailed explanations, but it does show how a problem can be worked through from beginning to the end. The principal aim is to develop deductive reasoning, which the student is then able to apply to unfamiliar situations, using as a basis a standard list of reactions and their associated mechanisms. This book is intended for First and Second Year students, but will not cover the fundamentals of the subject, which are more than adequately covered in a variety of texts already. A knowledge of electron accounting, such as the octet rule and Lewis structures, and the meaning of curly arrows and arrow pushing, is therefore assumed. However, this book will aim to reinforce and develop the use of these concepts by application of a generalised strategy to specific problems; this will be done using short multi-step reaction schemes. Note that because the emphasis is on the strategy of problem-solving, only a limited range of problems will be covered, and no attempt has been made to achieve a comprehensive list of all reactions. The aim is to demonstrate that this strategy is applicable to a wide variety of situations, and therefore an exhaustive list of problems is considered to be inappropriate; in fact, this would defeat the very purpose of this book.

A novel layout is used, in which two facing pages will have the problem and answer. On the left page is the problem and the overall strategy; on the right side of this page are broad hints corresponding to each step. These hints are not intended to give a detailed explanation of the answer, but to provide a guide to the approach to arriving at the answer. The right hand page will have a complete worked solution. Placing a piece of A4 paper on the right hand page will both provide a working space and hide the full answer from view. The intention is that this will remove the temptation to look at the complete solution too early but still provide access both to the stepwise procedure for working through the problem and to the hints on the left page. Of course, maximum benefit from these problems will come only if they are worked through in their entirety before looking at the worked solution! The detailed answer will be as full as possible within the page constraints and will include, for example, proton transfers. A new innovation will be introduced regarding curly arrows; these are labelled in sequence thus (a,b,c,d), to clearly indicate to the student the starting point and the subsequent sequence of movement of electrons. It is hoped that the provision of both hints and worked solutions will cater for a variety of academic abilities.

However, it should be emphasised that no matter how well a strategy for problem-solving is developed, there is no substitute for a good knowledge of the subject. One might consider that learning organic chemistry is little different from learning a foreign language. The vocabulary of any language is very important and must be learnt, and for organic chemistry these are the standard reactions of the common functional groups. A checklist of the reactions which a student is expected to know, and which form the basis of the questions in that chapter, are summarised before each set of problems. Little detail is included, however, since there are many excellent texts available which cover the required material. Mechanisms, which might be considered to be the grammar of organic chemistry, are covered in considerable detail in this book. Experience shows that mechanisms are best learnt by repeated practice in problem-solving.

How to use the book

1. This text has been subdivided by functional groups, since this provides an instantly recognisable starting point, especially for the beginners. A key skill, therefore, which must be developed early, is the recognition of functional groups and recollection of their typical or characteristic reactions. This information is briefly summarised at the beginning of every chapter, but an important starting point is to prepare your own set of more detailed notes using your recommended texts and lecture notes as source material. There is no alternative but to commit this material to memory.

2. These characteristic reactions can be very often understood using some fundamental chemical principles; mechanisms provide a way of rationalising the conversion of starting materials to products. In order to devise plausible mechanisms (remember that the only way of verifying any postulate is by experiment), it is necessary first to be able to identify nucleophiles and electrophiles, Bronsted–Lowry and Lewis acids and bases, and leaving groups.

3. A further aid to problem-solving is to number the atoms in the starting material and the corresponding atoms in the product; this allows for effective atom 'book-keeping'.

4. In devising plausible mechanisms, it is usually most helpful to begin at an electron-rich centre (negative charge, lone pair, or carbon–carbon double or triple bond) and push electrons (i.e. begin an arrow thus: ⤻) from there.

5. Remember that a double-headed arrow (⤻) refers to movement of two electrons and a single headed arrow (⤳) to a single electron.

6. Each problem in this book is designed to illustrate a sequential strategy of thinking to solve a question of the type: 'Provide a plausible explanation of the following interconversion; in your answer, include mechanisms for each reaction step'. A typical example is:

Note that numbering of each reagent, but not the product, is already given at this stage. Also the sequence of reagents drawn means that the intermediate product of step (i) is subsequently treated with the reagents of step (ii), whose product in turn is treated with the reagents of step (iii) to give the final product shown.

7. The general strategy is:

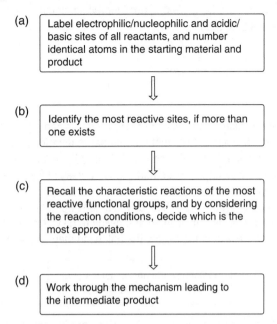

(a) Label electrophilic/nucleophilic and acidic/basic sites of all reactants, and number identical atoms in the starting material and product

(b) Identify the most reactive sites, if more than one exists

(c) Recall the characteristic reactions of the most reactive functional groups, and by considering the reaction conditions, decide which is the most appropriate

(d) Work through the mechanism leading to the intermediate product

The first step (box (a)) is an important preparatory stage, and this information is given on the right hand answer page above the dotted line, thus:

- -

Also included in this section are any preliminary reactions necessary to generate reactive species, for example:

$$PhCH_2NHCH_2CO_2H + Et_3N \rightleftharpoons PhCH_2NHCH_2CO_2^- + Et_3\overset{+}{N}H$$

- -

In order to solve a problem with multiple reaction sites, it is necessary to recognise which is the most reactive (box (b)). For the example shown, alkyl halides are electrophilic at the carbon adjacent to the halogen and alcohols are nucleophiles. Once the most reactive functional group has been identified, recollection of its characteristic reactions (box (c)) is a useful next step. In the example given, alkyl halides readily undergo nucleophilic substitution reactions. Once this is established, the next step (box (d)) can be deduced. To continue with this example, nucleophilic attack of thiolate, a potent nucleophile, at the alkyl halide generates the substitution product.

Fragments which are lost in the course of these steps (e.g. leaving groups, such as H_2O or Cl^-) are indicated as '–X' under the relevant reaction arrow, thus:

Two points are noteworthy: where possible, a series of unimolecular or bimolecular steps have been used; termolecular steps might make for a more concise answer but are kinetically much less favourable! It is important to remember that many reactions are in fact at equilibria, and that the overall transformation of starting materials to products often crucially depends on one or more irreversible steps in a reaction sequence; where space permits, this is indicated.

8. This process can be applied for as many iterations as are necessary in any given problem; when you have come to the end of one iterative cycle, the product (which could be an intermediate one or the final one of the question) is in a box, thus:

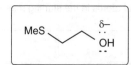

Once you have reached this stage, you will need to return to the beginning of the cycle (i.e. box (a)) and proceed through the sequence again. Note that the number of iterations required to reach the final product is different for each question; you will need to use your judgement accordingly. However, for a multi-step sequence like that in the example, you could expect to need at least the same number of iterations as there are steps, that is, in this case, three.

9. However, often a penultimate product is obtained which does not look like the desired product, but is in fact very close to it; this can be very misleading and needs to be watched for with care. Tautomerisation is a good example. Under these circumstances, the following step will be necessary:

> Recognise that this is not the final
> product, but is closely related to it

Note that only general acidic or basic work-up conditions are indicated, and this implies that the final product can be obtained by protonation or deprotonation respectively. If base or acid reagents are specified for any reaction sequence, a reaction (other than simple protonation or deprotonation) is implied.

10. All going well, you should now be in a position to:

> Write down the structure
> of the final product

11. Hints are provided adjacent to this general strategy; however, you may not need to use them, and if not, so much the better! Italicised terms should be familiar, but if not they can be checked in any suitable textbook.

12. The detailed answer is provided on the right hand page, allowing you to check your answer. Avoid the temptation to look at this until you have entirely finished the question!

13. There are 15 questions per chapter, designed to cover as many of the full range of characteristic reactions for each functional group as possible. On the website associated with the book, there are supplementary questions which are designed to reinforce the lessons of each detailed question.

14. Remember that this strategy is an aid to solving problems and is not always universally applicable; all problems are different, and slavish following of this approach is no guarantee to certain success. This strategy is no substitute for thinking!

15. There is one particularly important limitation to this strategy; it is designed specifically for mechanisms involving polar intermediates (hence the emphasis on electrophilic and nucleophilic processes). The strategy is not, however, directly applicable to radical reactions or pericyclic reactions, and reactions of this type have therefore been largely, although not exclusively, omitted. In any case, these types of reaction are considered to be too advanced for introductory organic chemistry.

16. Obviously, this approach is designed to develop your understanding of the subject, and in the short term, to be of use in those all-important examinations. The strategy is meant to make problem solving easier, and even fun! Enjoy them!

1 Nucleophilic substitution and elimination

Nucleophilic substitution: *S*$_N$*1 and S*$_N$*2 reactions*

- S$_N$1 is stepwise and unimolecular, proceeding through an intermediate carbocation, with a rate equation of Rate α [R^{3}R^{2}R^{1}CX], that is, proportional only to the concentration of alkyl halide starting material. The order of stability of the carbocation depends on structure, R$_3$C$^+$ > R$_2$CH$^+$ > RCH$_2^+$ > CH$_3$C$^+$ > H$_3$C$^+$, and rearrangements, by either hydrogen or carbon migrations, are possible.
- S$_N$2 is bimolecular with simultaneous bond-making and bond-breaking steps, but does not proceed through an intermediate, with a rate equation of Rate α [R^{3}R^{2}R^{1}CX][Nu$^-$], that is, the rate is reaction is proportional to both the concentration of alkyl halide starting material and the nucleophile.
- The nature of the substrate structure, nucleophile, leaving group, and solvent polarity can all alter the mechanistic course of the substitution.
- There are important stereochemical consequences of the S$_N$1 and S$_N$2 mechanisms (the former proceeds with racemisation and the latter with inversion).
- Steric effects are particularly important in the S$_N$2 reaction (neopentyl halides are unreactive).
- Neighbouring group participation in S$_N$1 reactions can be important.
- Special cases: (i) Allylic nucleophilic displacement: S$_N$1′ and S$_N$2′; (ii) Aryl (PhX) and vinylic (R$_2$C = CRX) halides: these are generally unreactive towards nucleophilic displacement, although benzylic (PhCH$_2$X) and allylic (RCH = CHCH$_2$X) are more reactive.

Elimination: E$_1$ and E$_2$ eliminations

- E$_1$ is stepwise and unimolecular, proceeding through an intermediate carbocation; E$_2$ is bimolecular with simultaneous bond-making and bond-breaking steps but does not proceed through an intermediate.
- The Saytzev's Rule and Hofmann's Rule can be used to predict the orientation of elimination, and the stereochemistry is preferentially antiperiplanar.
- Elimination and substitution are often competing reactions.

How to Solve Organic Reaction Mechanisms: A Stepwise Approach, First Edition. Mark G. Moloney.
© 2015 John Wiley & Sons, Ltd. Published 2015 by John Wiley & Sons, Ltd. Companion website: www.wiley.com/go/moloney/mechanisms

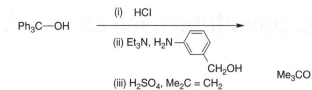

Label electrophilic/nucleophilic and acidic/basic sites of all reactants, and number identical atoms in the starting material and product.	Alcohols are nucleophiles and bases, since the oxygen possesses lone pairs. HCl is a strong acid and fully ionised (pKa = –7). Triethylamine is a weak base and sulfuric acid is also a very strong acid.
Identify the most reactive sites, if more than one exists.	Aromatic rings can be protonated, but the alcohol is the most basic and nucleophilic site of Ph_3COH.
Recall the characteristic reactions of the most reactive functional groups; and by considering the reaction conditions, decide which is the most appropriate.	Alcohols are easily protonated by strong acids, which converts the hydroxyl into a good *leaving group*, an *oxonium* ion, which is able to depart as water, leaving a carbocation intermediate.
Work through the mechanism leading to the intermediate product.	The leaving group departs to give a tertiary carbocation, which is also *resonance* stabilised, called the triphenylmethyl cation, which is then intercepted by chloride.
Repeat the above four steps.	* Triphenylmethyl chloride readily undergoes S_N1 reactions; departure of the good *leaving group* (chloride) regenerates the triphenylmethyl carbocation, which is intercepted by the most nucleophilic functional group of the aniline reagent, that is, the amine group. A series of proton transfers then gives the product. * Under stongly acidic conditions (H_2SO_4), isobutene is protonated (*Markovnikov* addition) to give a *t*-butyl cation; this is intercepted to give the ether product in its protonated form.
Recognise that this is not the final product, but is closely related to it.	Deprotonation of this *oxonium* cation gives the ether product.
Write down the structure of the final product.	

$$Nu^{\ominus} \; + \; R\text{-}X \; \longrightarrow \; R\text{-}Nu \; + \; X^{\ominus}$$

Now try questions 1.8 and 1.9

1.2

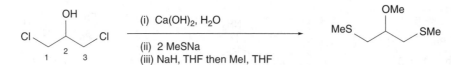

(i) Ca(OH)$_2$, H$_2$O

(ii) 2 MeSNa
(iii) NaH, THF then MeI, THF

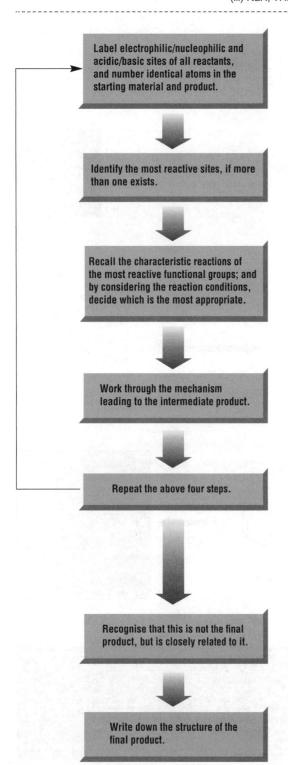

Label electrophilic/nucleophilic and acidic/basic sites of all reactants, and number identical atoms in the starting material and product.

Alkyl chlorides are good electrophiles (chloride is a good leaving group and its electronegativity polarises the C-Cl bond). Alcohols are nucleophiles and weak acids. Calcium hydroxide is a weak base.

Identify the most reactive sites, if more than one exists.

Under the basic conditions of the reaction, the alcohol function is deprotonated to give an alkoxide anion which is very nucleophilic. The alkyl chloride is the most electrophilic site.

Recall the characteristic reactions of the most reactive functional groups; and by considering the reaction conditions, decide which is the most appropriate.

Alkyl halides readily undergo nucleophilic substitution reactions with alkoxides to give ethers (the *Williamson ether synthesis*). In this case, the reaction would be an intramolecular one.

Work through the mechanism leading to the intermediate product.

The alkoxide undergoes an intramolecular *nucleophilic substitution* reaction with the alkyl chloride to give an epoxide.

Repeat the above four steps.

* Methanethiolate is a good nucleophile, and attacks both the C-Cl bond and the epoxide function (which has two electrophilic sites) at the less hindered end, to give the alkoxide product; this is protonated on work-up.

* Sodium hydride is a good base, and deprotonates the alcohol; *alkylation* with MeI, via a *nucleophilic substitution* mechanism, gives the final ether product (Williamson ether synthesis).

Recognise that this is not the final product, but is closely related to it.

Not needed here.

Write down the structure of the final product.

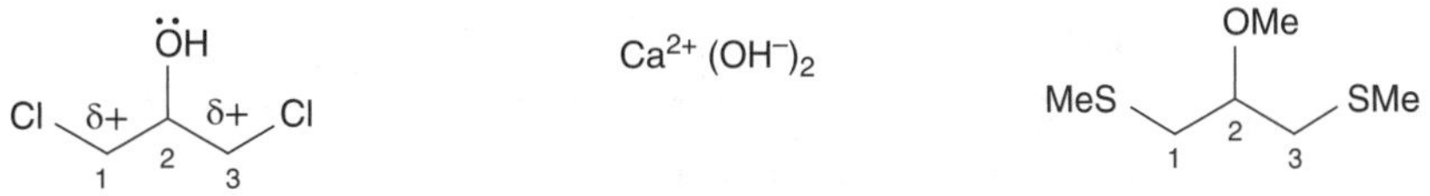

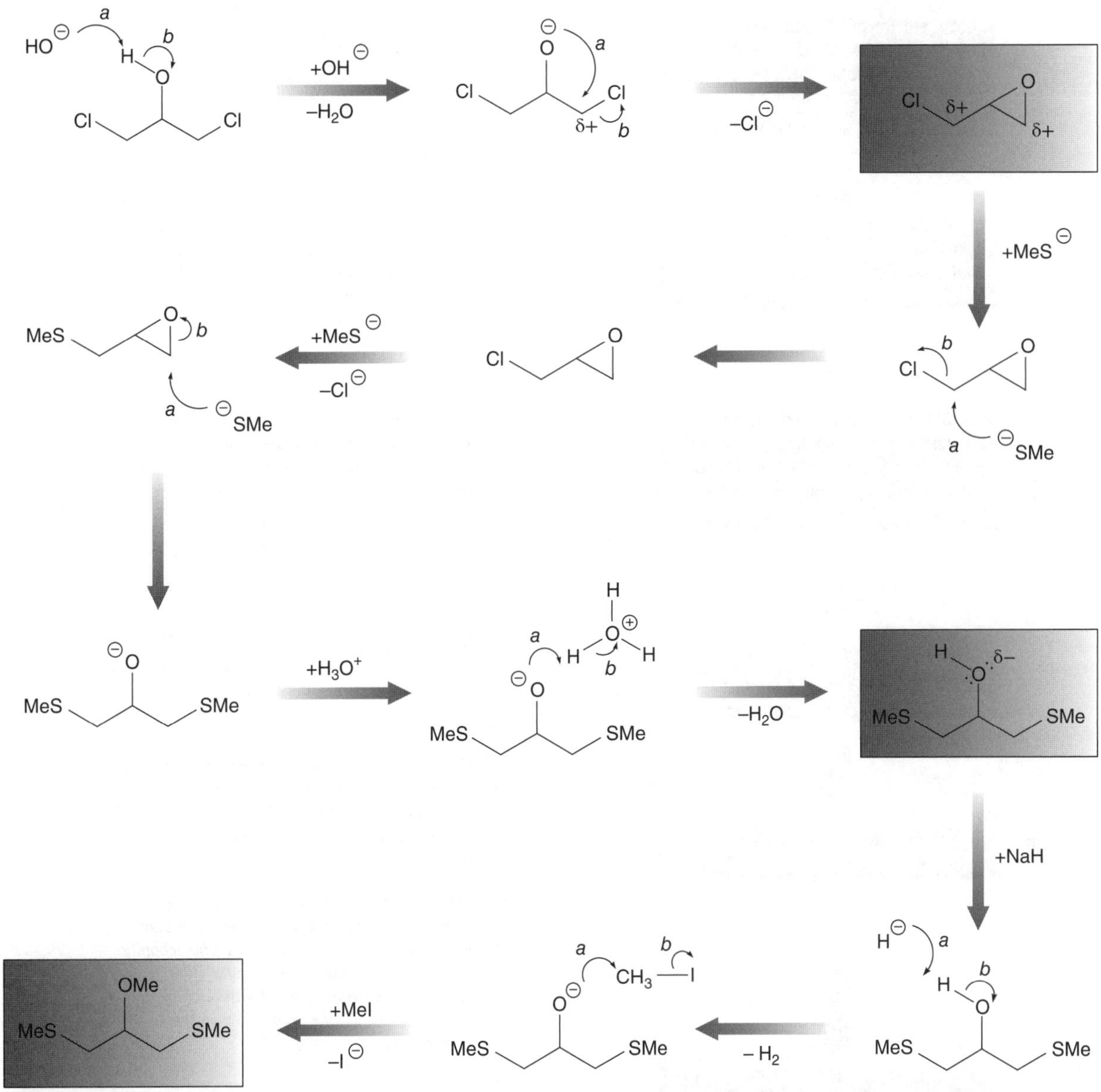

Cl $\overset{\delta+}{\underset{1}{}}$ $\overset{\ddot{O}H}{\underset{2}{}}$ $\overset{\delta+}{\underset{3}{}}$ Cl

$Ca^{2+} (OH^-)_2$

$\overset{OMe}{\underset{2}{}}$
MeS $\underset{1}{}$ SMe $\underset{3}{}$

HO⁻ a b H O ... Cl Cl

+OH⁻
−H_2O

O⁻ a ... Cl Cl $\delta+$ b

−Cl⁻

Cl $\delta+$ O $\delta+$

+MeS⁻

Cl O b a ⁻SMe

Cl O

+MeS⁻
−Cl⁻

MeS O b a ⁻SMe

MeS O⁻ SMe

+H_3O^+

H O⁺ H H a b O⁻ MeS SMe

−H_2O

H O $\delta-$ MeS SMe

+NaH

H⁻ a H O b MeS SMe

−H_2

O⁻ MeS SMe a CH_3 b I

+MeI
−I⁻

OMe MeS SMe

Summary: This question involves several examples of nucleophilic substitution (S_N2) reactions:

Nu⁻ + R-X ⟶ R-Nu + X⁻

Now try questions 1.10 and 1.16

1.3

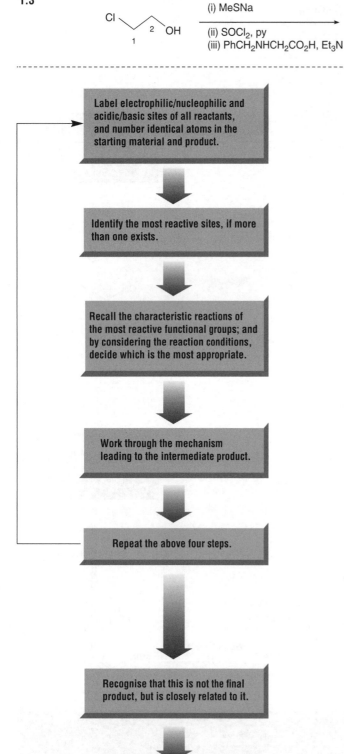

(i) MeSNa

(ii) SOCl₂, py
(iii) PhCH₂NHCH₂CO₂H, Et₃N

Label electrophilic/nucleophilic and acidic/basic sites of all reactants, and number identical atoms in the starting material and product.

Alkyl chlorides are good electrophiles (chloride is a good leaving group) and alcohols are good nucleophiles (the oxygen has two lone pairs). Methanethiolate is an excellent nucleophile (the sulfur is very polarisable and carries a negative charge).

Identify the most reactive sites, if more than one exists.

Since the reaction is with the highly nucleophilic reagent, MeS⁻, the most reactive site is the alkyl chloride. An alcohol is not reactive with a nucleophilic reagent.

Recall the characteristic reactions of the most reactive functional groups; and by considering the reaction conditions, decide which is the most appropriate.

Alkyl chlorides readily undergo *nucleophilic substitution* reactions, since they possess a leaving group, and are electrophilic by virtue of the electronegative halogen substituent.

Work through the mechanism leading to the intermediate product.

Since the reaction here is between a 1° alkyl chloride and a highly nucleophilic thiolate anion, an S_N2 mechanism is most likely.

Repeat the above four steps.

* Thionyl chloride is highly electrophilic, and converts the alcohol to the corresponding alkyl chloride *via* an *addition-elimination* process (with *neighbouring group* or *anchimeric* assistance of the SMe group).

* The nucleophilic hydroxyl oxygen of the carboxylic anion, generated by deprotonation of the carboxylic acid, undergoes a *nucleophilic substitution* reaction with the alkyl chloride formed in the previous step (with *anchimeric* assistance of the SMe group) to give the ester product.

Recognise that this is not the final product, but is closely related to it.

Not needed here.

Write down the structure of the final product.

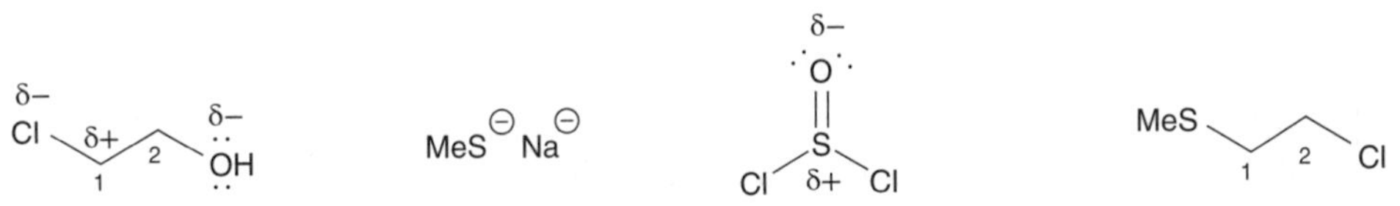

PhCH$_2$NHCH$_2$CO$_2$H + Et$_3$N ⇌ PhCH$_2$NHCH$_2$CO$_2^{\ominus}$ + Et$_3$NH$^{\oplus}$

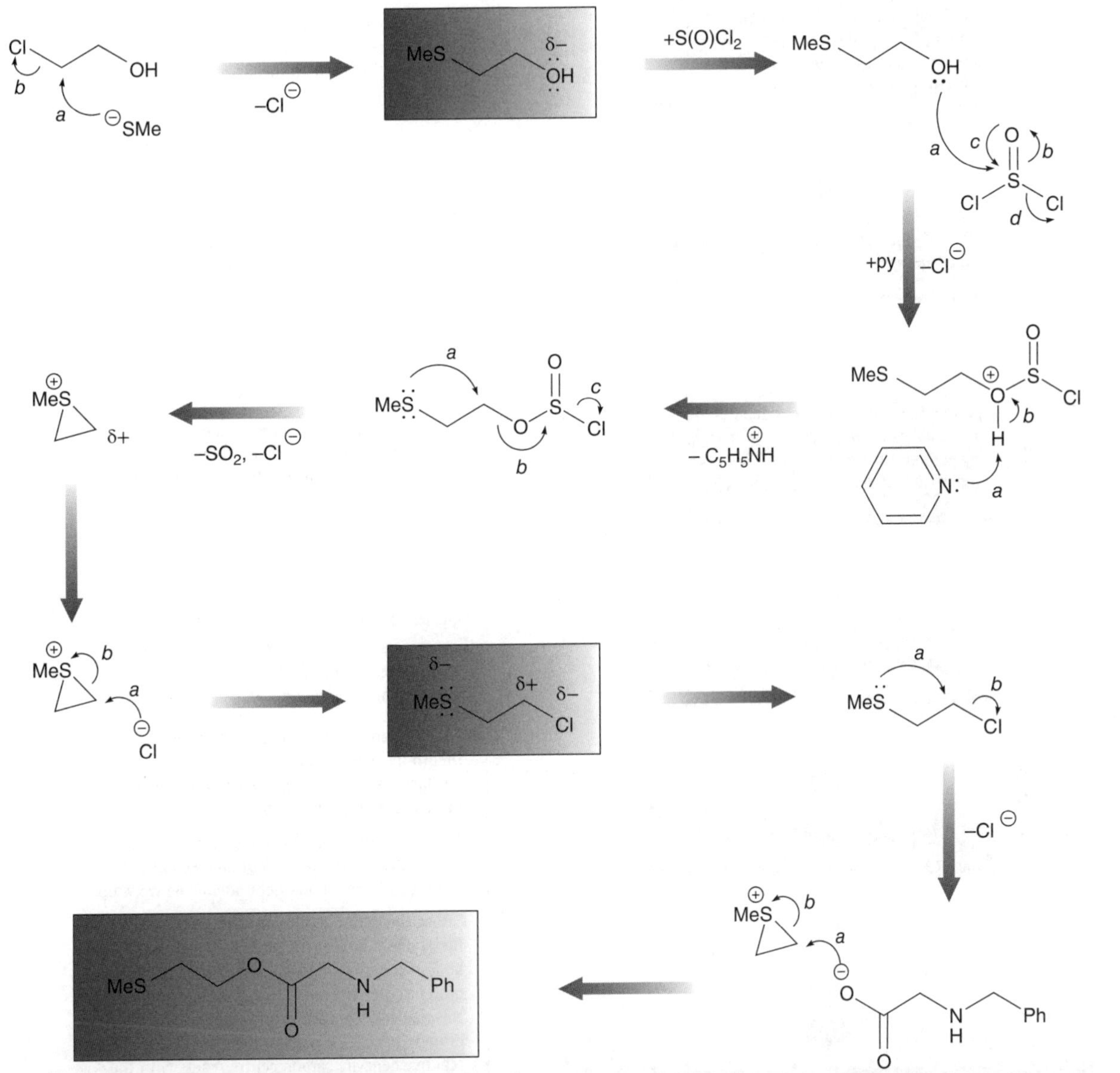

MeS$^{\ominus}$ Na$^{\oplus}$

+S(O)Cl$_2$

+py −Cl$^{\ominus}$

− C$_5$H$_5$NH$^{\oplus}$

−SO$_2$, −Cl$^{\ominus}$

−Cl$^{\ominus}$

Summary: This is an example of *anchimeric assistance* (also called *neighbouring group participation*) in nucleophilic substitution reactions

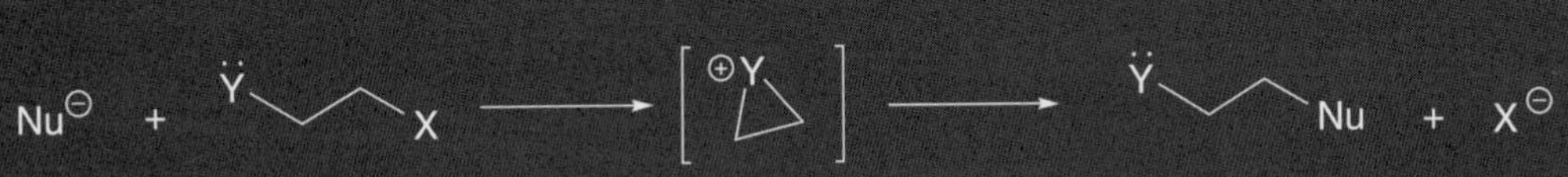

Nu$^{\ominus}$ + Ÿ⁀X ⟶ [$^{\oplus}$Y△] ⟶ Ÿ⁀Nu + X$^{\ominus}$

Now try questions 1.11 and 1.17

1.4

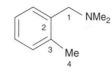

(i) EtBr, EtOH, Δ′
(ii) NaOAc, AcOH, Δ′

(iii) NaOH, Δ′, then acidic work-up
(iv) 2 BuLi, then EtBr (1 equiv.),
     then acidic work-up

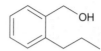

Label electrophilic/nucleophilic and acidic/basic sites of all reactants, and number identical atoms in the starting material and product.

Amines are good nucleophiles (the nitrogen has a lone pair). Alkyl bromides are good electrophiles, since bromine is electronegative and bromide is a good leaving group.

Identify the most reactive sites, if more than one exists.

Although an aromatic ring is a possible nucleophile, the most nucleophilic site of this molecule is the dimethylamino group. The most reactive electrophile is ethyl bromide.

Recall the characteristic reactions of the most reactive functional groups; and by considering the reaction conditions, decide which is the most appropriate.

Alkyl halides readily undergo *nucleophilic substitution* reaction, and in this case; the nucleophile is the dimethylamino function.

Work through the mechanism leading to the intermediate product.

Since bromoethane is a 1° alkyl halide, and amines are good nucleophiles, the reaction will proceed by an S_N2 process, to give a quaternary ammonium salt.

Repeat the above four steps.

* Quaternary ammonium groups are good (neutral) leaving groups, and are easily displaced by the nucleophile acetate, giving in this case an acetate ester.

* Esters are easily hydrolysed under alkaline conditions (*addition-elimination mechanism*) to give the alcohol product upon acidic work-up.

* Reaction with butyllithium (a strong base) firstly deprotonates the more acidic alcohol, and only then deprotonates the benzylic methyl group to give a resonance stabilised carbanion; this nucleophilic carbanion is quenched with bromoethane (S_N2 reaction), but only the most reactive benzylic carbon centre reacts with the one equivalent of EtBr.

Recognise that this is not the final product, but is closely related to it.

The alkoxide is protonated on acidic work-up to give the alcohol product.

Write down the structure of the final product.

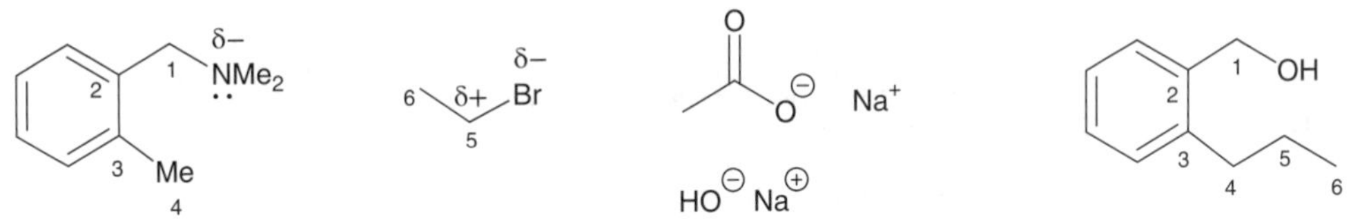

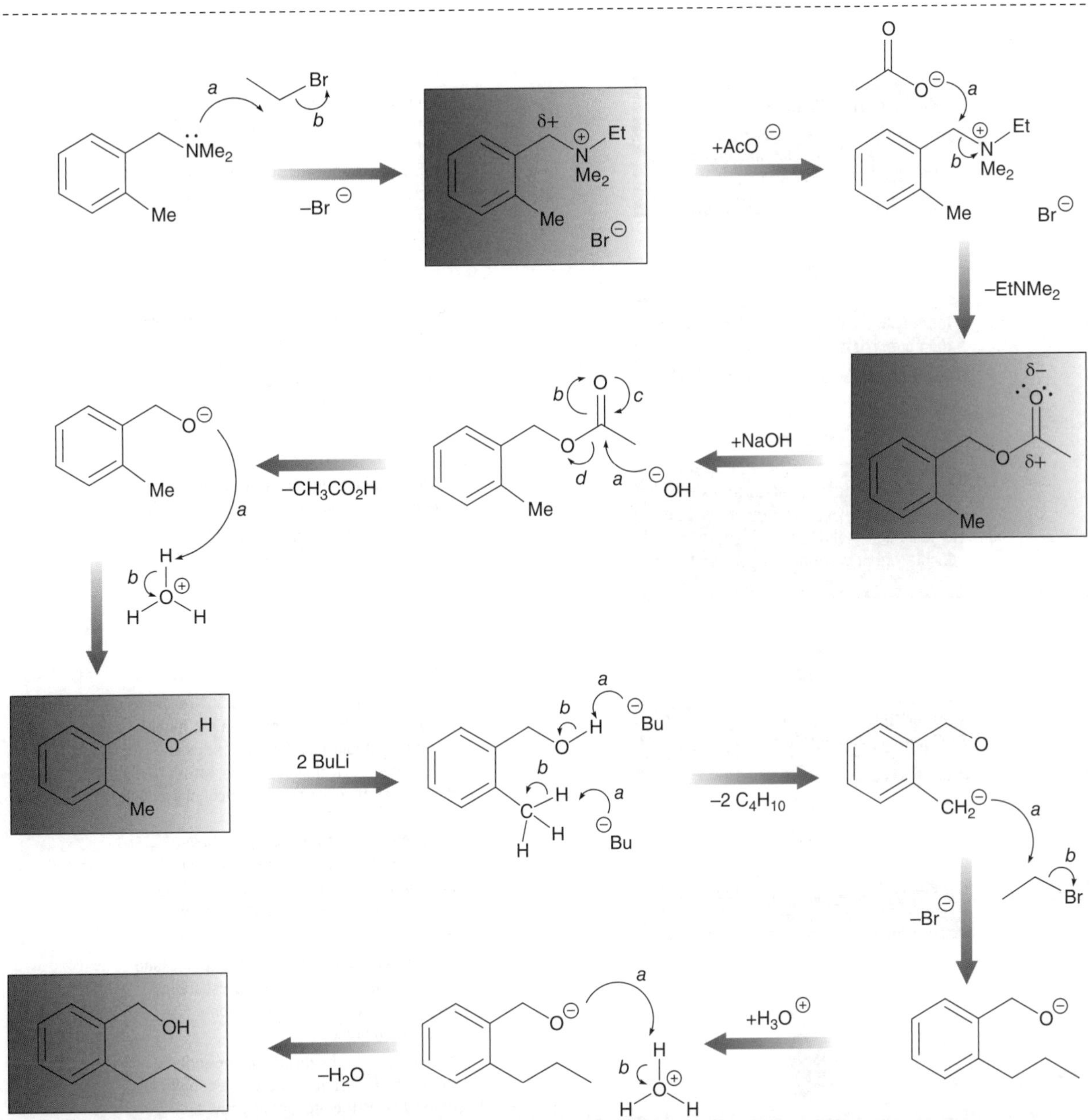

$\delta-$
1 NMe$_2$
2
3 Me
4

$\delta-$
6 $\delta+$ Br
5

O
O$^{\ominus}$ Na$^+$

HO$^{\ominus}$ Na$^{\oplus}$

1 OH
2
3 5
4 6

---

a
Br
b

NMe$_2$

Me

$-$Br$^{\ominus}$

$\delta+$
N$^{\oplus}$ Et
Me$_2$

Me

Br$^{\ominus}$

+AcO$^{\ominus}$

O
O$^{\ominus}$ *a*
N$^{\oplus}$ Et
b Me$_2$

Me

Br$^{\ominus}$

$-$EtNMe$_2$

$\delta-$
O
O
$\delta+$

Me

+NaOH

b O *c*
O
d O *a*
$^{\ominus}$OH

Me

O$^{\ominus}$

Me

a
$-$CH$_3$CO$_2$H

H
b O$^{\oplus}$
H H

O H

Me

2 BuLi

a
b H $^{\ominus}$ Bu
O
b
H *a*
C $^{\ominus}$
H Bu
H

$-$2 C$_4$H$_{10}$

O

CH$_2$$^{\ominus}$ *a*

b
Br

$-$Br$^{\ominus}$

O$^{\ominus}$

+H$_3$O$^{\oplus}$

a
O$^{\ominus}$
b H
O$^{\oplus}$
H H

OH

$-$H$_2$O

Summary: This question includes an example of nucleophilic attack at a benzylic position:

Ph X + Nu$^{\ominus}$ $\longrightarrow$ Ph Nu + X$^{\ominus}$

Now try questions 1.12 and 1.18

9

1.5

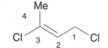 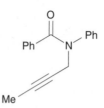

(i) Na_2CO_3, H_2O
(ii) 2 $NaNH_2$, NH_3, then acidic work-up
(iii) $SOCl_2$, py
(iv) $PhC(O)N^-Ph$ Na^+

Label electrophilic/nucleophilic and acidic/basic sites of all reactants, and number identical atoms in the starting material and product.

Alkyl chlorides are electrophiles, since chlorine is electronegative and chloride is a good leaving group. Sodium carbonate is a weak base; in aqueous solutions, hydroxide is generated which is a good base as well as a good nucleophile.

Identify the most reactive sites, if more than one exists.

The *allylic* chloride is the most electrophilic site (*vinylic* chlorides have a much stronger C–Cl bond and are not electrophilic). Hydroxide is the only available nucleophile.

Recall the characteristic reactions of the most reactive functional groups; and by considering the reaction conditions, decide which is the most appropriate.

Allylic chlorides are very susceptible to *nucleophilic substitution* reactions, which can be either by an S_N1 or S_N2 mechanism, depending on the substrate and solvent.

Work through the mechanism leading to the intermediate product.

Allylic chlorides easily undergo S_N1 reactions in polar solvents, proceeding via the highly *resonance* stabilised allyl cation; this is intercepted by hydroxide, which, after deprotonation, gives the allyl alcohol product.

Repeat the above four steps.

* Sodamide is a strong base; the first equivalent first deprotonates the alcohol function; the second then induces *elimination* of the vinylic chloride to give the alkyne product.

* Thionyl chloride is highly electrophilic, and converts the alcohol to the corresponding propargyl chloride *via* an *addition-elimination* process.

* *Nucleophilic substitution* by the sodium salt of PhC(O)NHPh (N more nucleophilic than O) gives the product directly.

Recognise that this is not the final product, but is closely related to it.

Not needed here.

Write down the structure of the final product.

Summary: This question gives more examples of nucleophilic substitution and elimination reactions:

Now try questions 1.13 and 1.19

1.6

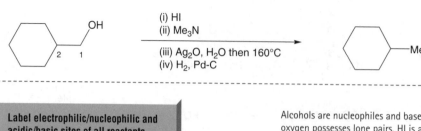

(i) HI
(ii) Me$_3$N
(iii) Ag$_2$O, H$_2$O then 160°C
(iv) H$_2$, Pd-C

Label electrophilic/nucleophilic and acidic/basic sites of all reactants, and number identical atoms in the starting material and product.

Alcohols are nucleophiles and bases, since the oxygen possesses lone pairs. HI is a strong acid, and fully ionised (pK$_a$ = −10).

Identify the most reactive sites, if more than one exists.

The alcohol is the only basic and nucleophilic site in this molecule, and HI the most electrophilic reagent.

Recall the characteristic reactions of the most reactive functional groups; and by considering the reaction conditions, decide which is the most appropriate.

Alcohols are readily protonated by acids, thereby converting the hydroxyl group into an *oxonium* ion, which is able to depart as water; they therefore readily undergo *nucleophilic substitution* reactions under acidic conditions with suitable nucleophiles.

Work through the mechanism leading to the intermediate product.

The *oxonium* ion is a good leaving group, and is easily displaced by the good nucleophile, iodide, in an S$_N$2 process (1° substrate, good nucleophile).

Repeat the above four steps.

* Trimethylamine is a good base and nucleophile. Alkyl iodides are also reactive to S$_N$2 reactions, and the iodide is easily displaced to give a quaternary ammonium iodide.

* Treatment with silver oxide converts the iodide salt to the hydroxide salt (driven by the precipitation of AgI); heating of this salt causes an *elimination* reaction (*Hofmann elimination*) to give the corresponding alkene.

* Hydrogenation of the alkene using Pd supported on charcoal as catalyst gives the alkane (*syn-* addition of H$_2$).

Recognise that this is not the final product, but is closely related to it.

Not needed here.

Write down the structure of the final product.

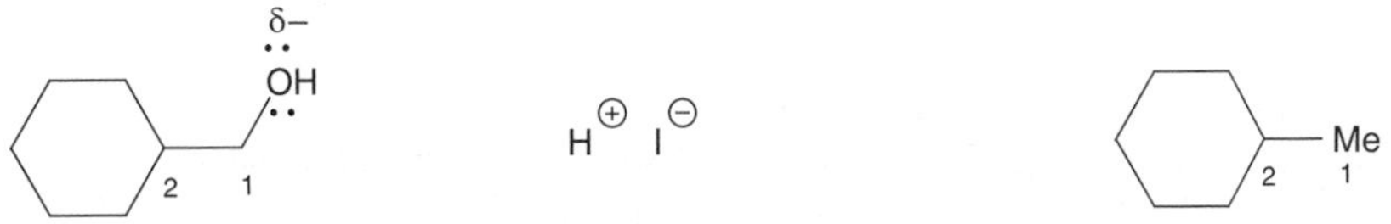

δ−

OH

2 1

H ⊕ I ⊖

Me

2 1

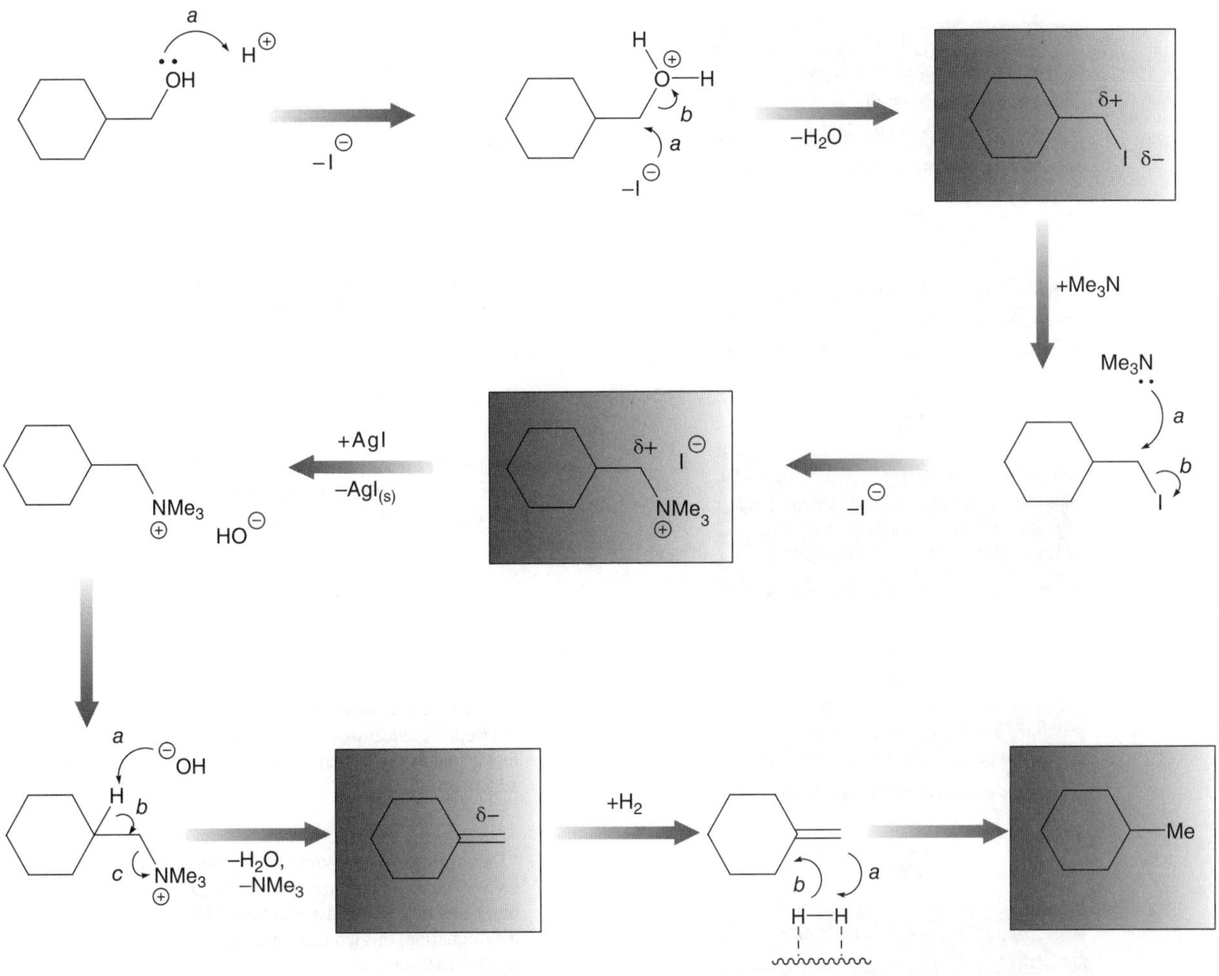

---

a

H ⊕

OH

−I ⊖

H
O ⊕ H
b
a
−I ⊖

−H₂O

δ+
I δ−

+Me₃N

Me₃N
a
b
I

−I ⊖

δ+ I ⊖

NMe₃ ⊕

+AgI
−AgI₍ₛ₎

NMe₃ ⊕ HO ⊖

a ⊖OH
H b
c NMe₃ ⊕

−H₂O,
−NMe₃

δ−

+H₂

b a

H—H

Me

Summary: This is an example of the Hofmann elimination reaction of quaternary ammonium iodides:

N⁺—
I−

Now try questions 1.14 and 1.20

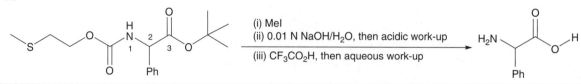

(i) MeI
(ii) 0.01 N NaOH/H$_2$O, then acidic work-up
(iii) CF$_3$CO$_2$H, then aqueous work-up

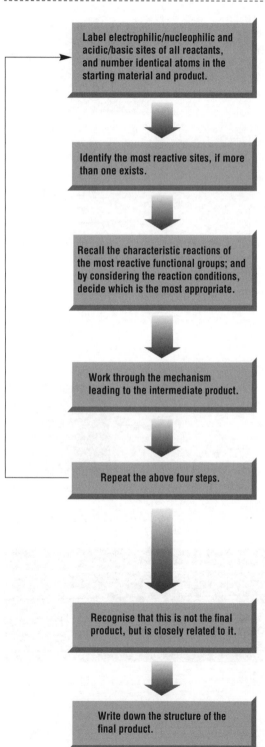

Label electrophilic/nucleophilic and acidic/basic sites of all reactants, and number identical atoms in the starting material and product.	Sulfur, oxygen and nitrogen are all nucleophilic, since all possess lone pairs. Methyl iodide is a good electrophile, since iodine is electronegative and iodide is a good leaving group.
Identify the most reactive sites, if more than one exists.	Sulfur is the most nucleophilic heteroatom, since it is the least electronegative of O, N and S. Methyl iodide is the most reactive electrophile.
Recall the characteristic reactions of the most reactive functional groups; and by considering the reaction conditions, decide which is the most appropriate.	Alkyl halides readily undergo *nucleophilic substitution* reactions, the nucleophile in this case being the sulfur atom.
Work through the mechanism leading to the intermediate product.	The nucleophilic sulfur undergoes a *nucleophilic substitution* reaction with methyl iodide (S$_N$2) to generate a *sulfonium* cation.
Repeat the above four steps.	* The α-protons of the sulfonium cation are very acidic (highly stabilised conjugate base), and only weak base is required for deprotonation; this induces an *elimination* (E1CB) reaction. * Esters can be protonated on the carbonyl oxygen by strong acids (e.g. CF$_3$CO$_2$H); the ester alkyl group departs in an E$_1$ process, to give the amino acid product and a *t*-butyl cation. The cation is intercepted by water, to give an *oxonium* cation.
Recognise that this is not the final product, but is closely related to it.	Deprotonation of the *oxonium* cation on work-up gives the product, *t*-butyl alcohol.
Write down the structure of the final product.	

Summary: This is an example of a base catalysed β–elimination and acid-catalysed ester hydrolysis.

Now try questions 1.15, 1.21 and 1.9

1.8

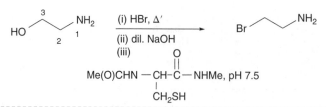

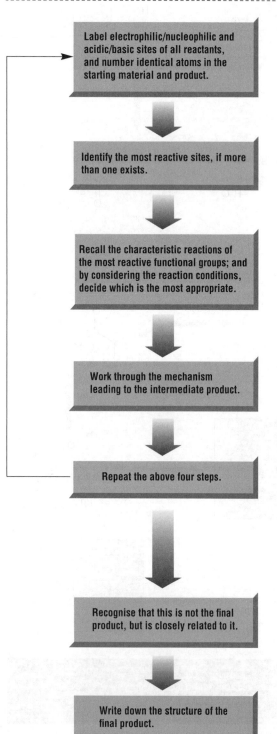

Label electrophilic/nucleophilic and acidic/basic sites of all reactants, and number identical atoms in the starting material and product.	Amines and alcohols are both basic, since they both possess non-bonded electron pairs. HBr is a strong acid, and fully ionised ($pK_a = -8$).
Identify the most reactive sites, if more than one exists.	Although amines and alcohols are both basic, amines are more so (nitrogen is less electronegative than oxygen). HBr is the strongest acid present.
Recall the characteristic reactions of the most reactive functional groups; and by considering the reaction conditions, decide which is the most appropriate.	Amines are easily protonated by hydrogen bromide, but so too are alcohols.
Work through the mechanism leading to the intermediate product.	The amine function is initially protonated by the hydrogen bromide. Protonation of the amine generates bromide anion.
Repeat the above four steps.	* HBr is a strong acid, easily capable of protonating the remaining alcohol. This converts the alcohol into an excellent hydroxonium leaving group, which can depart as water, if a suitable nucleophile attacks, which in this case would be bromide anion. Deprotonation on basic work-up generates the amine product. * Reaction with a cysteine derivative occurs by attack of the most nucleophilic sulfur atom on the bromide (with *anchimeric* assistance of the NH₂ group) to give the protonated form of the thioether product.
Recognise that this is not the final product, but is closely related to it.	Deprotonation on work-up generates the thioether product.
Write down the structure of the final product.	

16

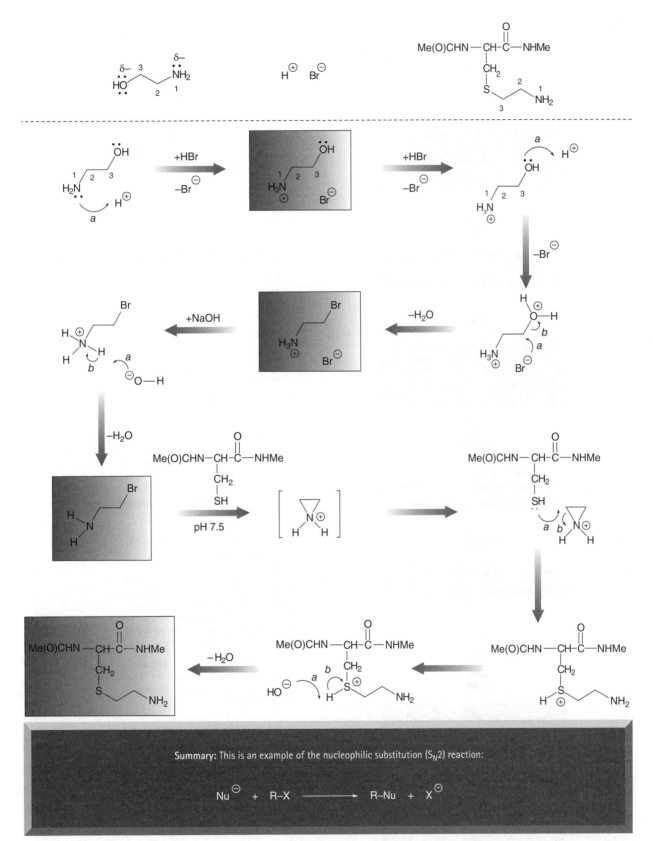

Summary: This is an example of the nucleophilic substitution (S_N2) reaction:

$$Nu^{\ominus} \ + \ R\!-\!X \ \longrightarrow \ R\!-\!Nu \ + \ X^{\ominus}$$

Now try question 1.9

1.9

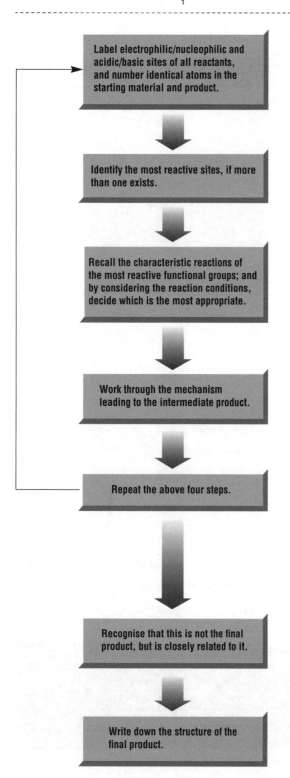

(i) HCl, Δ′
(ii) NaOPh

Cl $\diagup\diagdown\diagup\diagdown$ Cl

Label electrophilic/nucleophilic and acidic/basic sites of all reactants, and number identical atoms in the starting material and product.	Ethers have oxygen nucleophiles, since the oxygen possesses lone pairs, and HCl is a strong acid and fully ionised (pK_a = −7).
Identify the most reactive sites, if more than one exists.	The ether function is the only reactive group in this molecule, and HCl is the strongest acid.
Recall the characteristic reactions of the most reactive functional groups; and by considering the reaction conditions, decide which is the most appropriate.	Ethers are generally unreactive to most reagents, but may be cleaved by strong acids like HCl, to generate alkyl halides.
Work through the mechanism leading to the intermediate product.	The ether function can be protonated by the HCl, to generate an oxonium cation. This cation provides an excellent leaving group (water), which may be displaced by the weak nucleophile, chloride anion. This generates an alkyl halide product.
Repeat the above four steps.	* The alchoholic group simultaneously generated may react as before; protonation by the strong acid HCl gives the corresponding oxonium ion. This oxonium cation provides an excellent leaving group, which may be displaced by the weak nucleophile, chloride anion. This generates an alkyl halide product, and the dichloro product is thus formed. * Reaction of the dichloride with sodium phenoxide gives the product from double displacement of halogen by nucleophilic substitution.
Recognise that this is not the final product, but is closely related to it.	Not needed here.
Write down the structure of the final product.	

Summary: This question involves several examples of nucleophilic substitution (SN2) reactions:

$$Nu^{\ominus} + R\text{-}X \longrightarrow R\text{-}Nu + X^{\ominus}$$

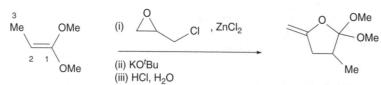

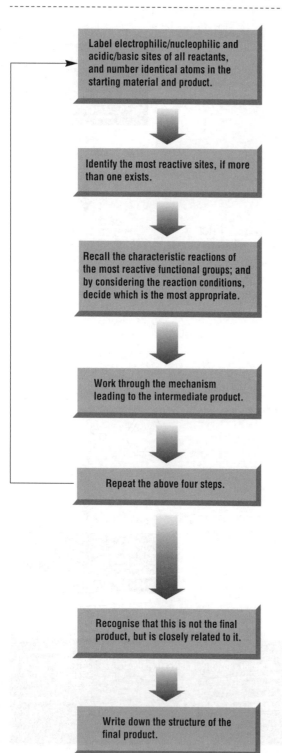

Label electrophilic/nucleophilic and acidic/basic sites of all reactants, and number identical atoms in the starting material and product.	Alkenes are good nucleophiles. Alkyl chlorides are electrophiles, since chlorine is electronegative and chloride is a good leaving group. Epoxides are also electrophiles, since oxygen is electronegative and can be a good leaving group. Zinc chloride is an excellent Lewis acid. Potassium *t*-butoxide is a strong base.
Identify the most reactive sites, if more than one exists.	Once activated by the Lewis acid, the epoxide is the most electrophilic site.
Recall the characteristic reactions of the most reactive functional groups; and by considering the reaction conditions, decide which is the most appropriate.	Epoxides are very susceptible to *nucleophilic substitution* reactions, especially when activated by Lewis acids. Under these circumstances, the alkyl chloride is the less reactive group. Alkenes are weak nucleophiles.
Work through the mechanism leading to the intermediate product.	The zinc cation co-ordinates to the epoxide, and this group is then attacked by the alkene. This attack is promoted by the presence two methoxy groups which activate the alkene, leading to the formation of an *oxonium cation*. Ring closure of the released alkoxide onto this oxonium cation generates the ring closed product.
Repeat the above four steps.	* Potassium *t*-butoxide is a strong base; this induces *elimination* of the alkyl chloride to give the alkene product. This is most likely to proceed by an E_2 mechanism. * Under aqueous acidic conditions, the acetal is readily *hydrolysed*, by the usual protonation–elimination–addition of water–elimination sequence for acetal hydrolysis.
Recognise that this is not the final product, but is closely related to it.	Not needed here.
Write down the structure of the final product.	

ZnCl$_2$

K$^{\oplus}$ ^{t}BuO$^{\ominus}$

a

+Zn^{2+}

b

a

a

b

−O^{t}Bu

a

b

c

− HO^{t}Bu

−Cl$^{\ominus}$

a

b

+H$_3$O$^{\oplus}$

−MeOH

a

b

+H$_2$O

−H$^+$ + H$^+$

a

b

−MeOH

Summary: This question involves nucleophilic substitution and elimination reactions:

Nu$^{\ominus}$ + R-X $\longrightarrow$ R-Nu + X$^{\ominus}$

Now try question 1.16

1.11

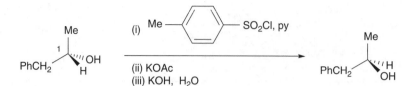

$$\text{PhCH}_2\overset{\overset{\displaystyle \text{Me}}{|}}{\underset{\underset{\displaystyle H}{|}}{\overset{1}{C}}}\text{OH} \quad \xrightarrow[\substack{\text{(ii) KOAc}\\\text{(iii) KOH, H}_2\text{O}}]{\text{(i)}} \quad \text{PhCH}_2\overset{\overset{\displaystyle \text{Me}}{|}}{\underset{\underset{\displaystyle \text{OH}}{|}}{\overset{}{C}}}\text{H}$$

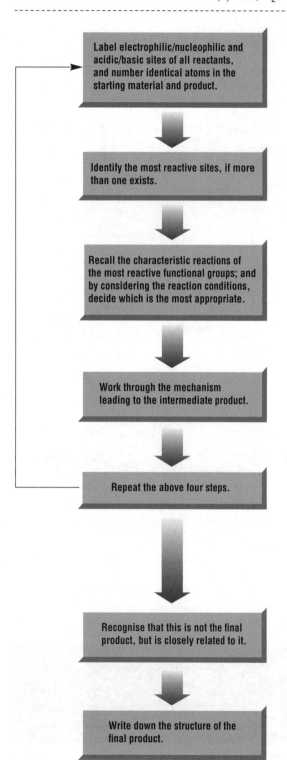

Label electrophilic/nucleophilic and acidic/basic sites of all reactants, and number identical atoms in the starting material and product.

Identify the most reactive sites, if more than one exists.

Recall the characteristic reactions of the most reactive functional groups; and by considering the reaction conditions, decide which is the most appropriate.

Work through the mechanism leading to the intermediate product.

Repeat the above four steps.

Recognise that this is not the final product, but is closely related to it.

Write down the structure of the final product.

Alcohols are good nucleophiles (oxygen possesses lone pairs). *p*-TsCl is an acid chloride derived from a sulfonic acid, and is electrophilic due to the electronegative oxygen atoms with a chloride being a good leaving group. Potassium acetate is a weak base and good nucleophile.

The alcohol is the only nucleophilic site, tosyl chloride is the only electrophile. Pyridine is a base.

Alcohols are easily converted to esters by acid chlorides, in this case a sulfonyl chloride.

The alcohol undergoes a *nucleophilic addition-elimination* reaction at the sulfonic acid group, with loss of chloride; this generates a protonated sulfonate ester. Deprotonation by pyridine generates the product.

* The tosylate group is an excellent leaving group, and this system is especially activated towards S_N2 reactions by attack from the weak nucleophile, acetate. Such reaction leads to displacement of the tosylate giving an ester product but with stereochemical *inversion*.

* Hydroxide is highly nucleophilic, and hydrolyses the ester; this converts the ester to the alcoholate anion via an *addition-elimination* process.

The basic conditions of this reaction gives an alcoholate product, which is reprotonated on acidic work-up.

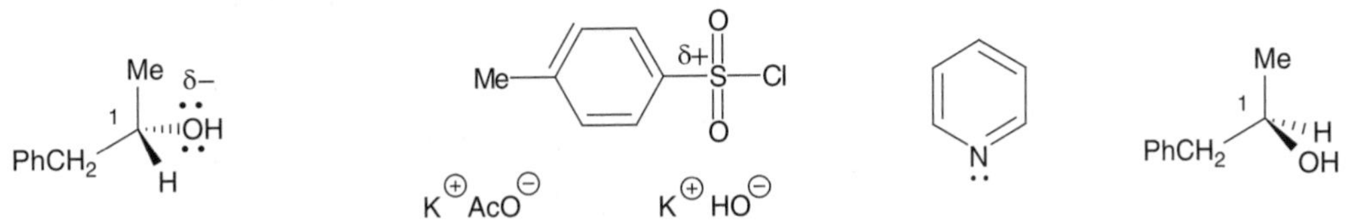

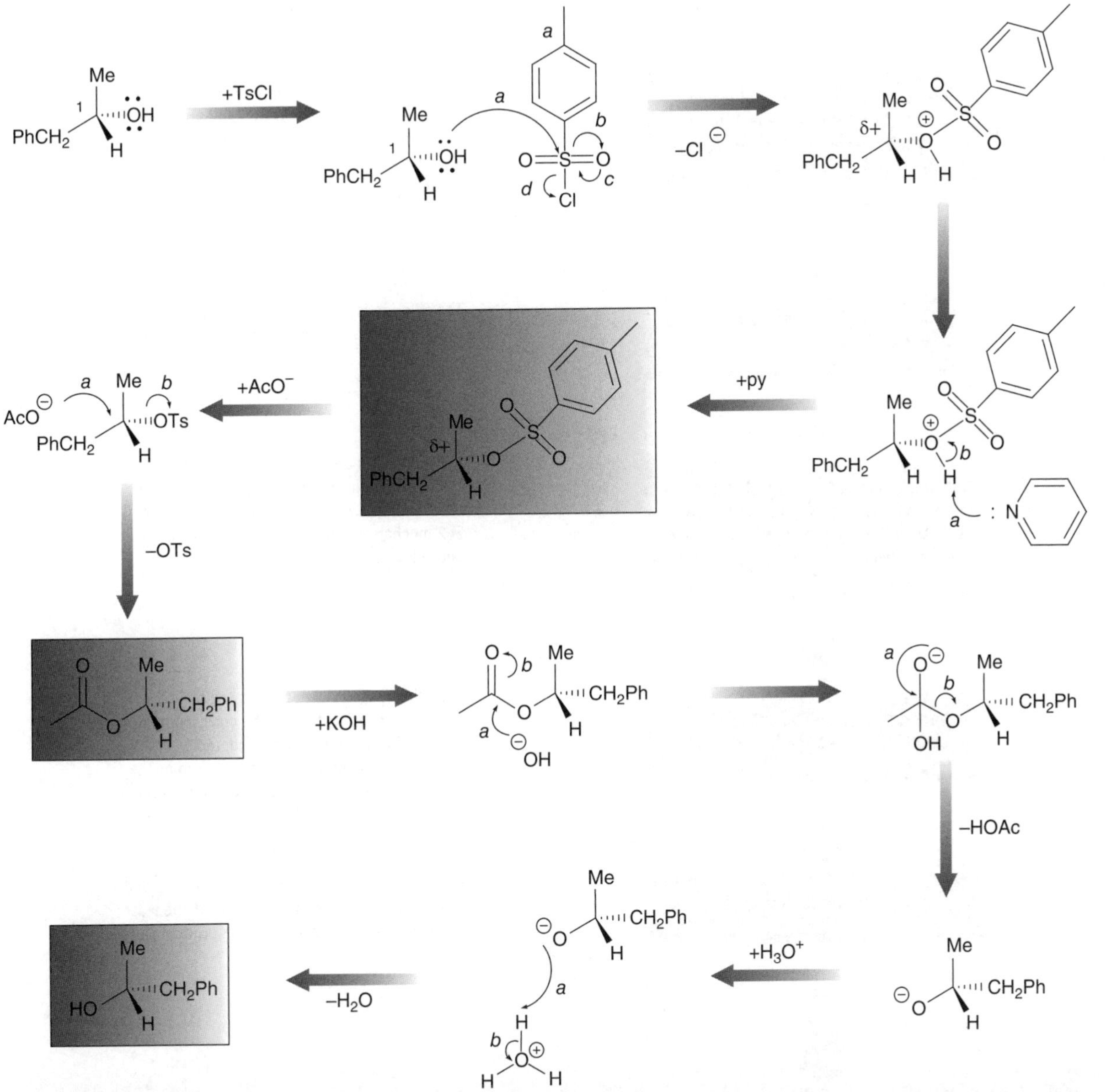

Me δ−
PhCH₂—¹C⁎⁎OH
H

Me
(p-tolyl)—S(=O)₂—Cl   δ+

K⁺ AcO⁻   K⁺ HO⁻

pyridine : N

Me
PhCH₂—¹C⁎⁎H
OH

+TsCl

−Cl⁻

a

+py

b

+AcO⁻

−OTs

+KOH

−HOAc

+H₃O⁺

−H₂O

Me
HO—C⁎⁎CH₂Ph
H

Summary: This question involves several examples of nucleophilic substitution (S_N2) reactions:

$$Nu^- + R\text{-}X \longrightarrow R\text{-}Nu + X^-$$

Now try question 1.17

1.12

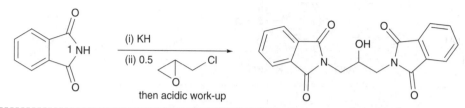

Label electrophilic/nucleophilic and acidic/basic sites of all reactants, and number identical atoms in the starting material and product.	Phthalimide is acidic, and potassium hydride is a strong base. The epoxide has two electrophilic sites, at the less hindered end of the epoxide, and at the chloride-bearing carbon.
Identify the most reactive sites, if more than one exists.	The chloride is the most electrophilic site, and the N–H is the only acidic bond present.
Recall the characteristic reactions of the most reactive functional groups; and by considering the reaction conditions, decide which is the most appropriate.	Alkyl chlorides are electrophiles and are very susceptible to *nucleophilic substitution* reactions, which can be either by an S_N1 or S_N2 mechanism, depending on the substrate and solvent.
Work through the mechanism leading to the intermediate product.	Phthalimide is easily deprotonated by potassium hydride, to give the corresponding anion. Alkyl chlorides easily undergo S_N2 reactions with good nucleophiles, such as the phthalimide anion, and give the amine product.
Repeat the above four steps.	Phthalimide anion is a strong nucleophile, and reacts further to open the epoxide at the least hindered end, to give an alkoxide product.
Recognise that this is not the final product, but is closely related to it.	The alkoxide product is protonated on acidic work-up to give the final product.
Write down the structure of the final product.	

Summary: This question involves several examples of nucleophilic substitution (S_N2) reactions:

$$Nu^{\ominus} \ + \ R\text{-}X \ \longrightarrow \ R\text{-}Nu \ + \ X^{\ominus}$$

Now try questions 1.18

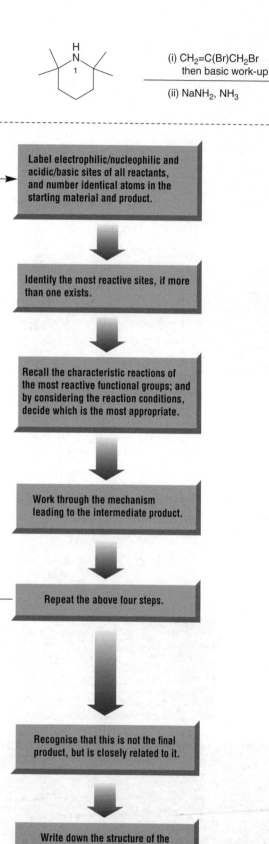

(i) CH$_2$=C(Br)CH$_2$Br
then basic work-up

(ii) NaNH$_2$, NH$_3$

Label electrophilic/nucleophilic and acidic/basic sites of all reactants, and number identical atoms in the starting material and product.	Allyl bromides are excellent electrophiles, since bromine is electronegative and bromide is a good leaving group. Amines are weak bases and good nucleophiles. Sodamide is an excellent base.
Identify the most reactive sites, if more than one exists.	The *allylic* bromide is the most electrophilic site (*vinylic* bromides have a much stronger C–Br bond) and the amine is the only nucleophile present.
Recall the characteristic reactions of the most reactive functional groups; and by considering the reaction conditions, decide which is the most appropriate.	Allylic bromides are very susceptible to *nucleophilic substitution* reactions, which can be either by an S$_N$1 or S$_N$2 mechanism, depending on the solvent.
Work through the mechanism leading to the intermediate product.	Allylic bromides easily undergo nucleophilic substitution reactions, even with weak amine nucleophiles; the initially formed product is deprotonated on work-up, to give the allylic amine product.
Repeat the above four steps.	Sodamide is a strong base; it induces *elimination* of the vinylic chloride to give the alkyne product.
Recognise that this is not the final product, but is closely related to it.	Not needed here.
Write down the structure of the final product.	

Summary: This question involves examples of nucleophilic substitution (S$_N$2) and elimination (E2) reactions:

$$Nu^{\ominus} + R\text{-}X \longrightarrow R\text{-}Nu + X^{\ominus}$$

Now try questions 1.19

1.14

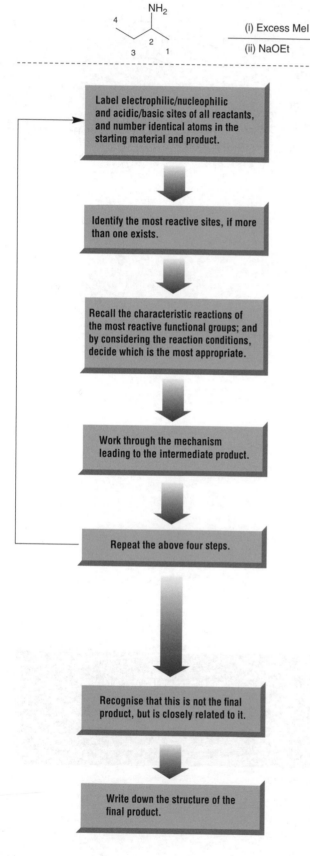

(i) Excess MeI
(ii) NaOEt

Label electrophilic/nucleophilic and acidic/basic sites of all reactants, and number identical atoms in the starting material and product.

Alkyl iodides are electrophiles, since iodine is electronegative and iodide is a good leaving group. Amines are good nucleophiles, since the nitrogen possesses a lone pair. Sodium ethoxide is an excellent base.

Identify the most reactive sites, if more than one exists.

The alkyl iodide is the only electrophilic site and the amine the only nucleophilic site.

Recall the characteristic reactions of the most reactive functional groups; and by considering the reaction conditions, decide which is the most appropriate.

Alkyl iodides are very susceptible to *nucleophilic substitution* reactions, which can be either by an S_N1 or S_N2 mechanism, depending on the substrate.

Work through the mechanism leading to the intermediate product.

Methyl iodide easily undergoes S_N2 reactions since it is very unhindered; amines are weak nucleophiles, and reaction gives the ammonium intermediate which is deprotonated to give the methylamine product.

Repeat the above four steps.

* This methylamine product is a weak base and good nucleophile, better in fact than the starting material as a result of inductive electron release of the N substituents; the alkylation reaction therefore repeats to give the dimethylamine.
* This dimethylamine product is an even better nucleophile; the alkylation reaction therefore repeats to give the trimethylammonium product. This is called *exhaustive* methylation.
* Sodium ethoxide then induces *elimination* of the trimethylammonium groups to give the alkene product, in a reaction which is controlled kinetically, to give the less substituted alkene (*Hofmann* elimination).

Recognise that this is not the final product, but is closely related to it.

Not needed here.

Write down the structure of the final product.

Summary: This is an example of the quarternisation of an amine followed by Hofmann elimination reaction:

Now try question 1.20

1.15

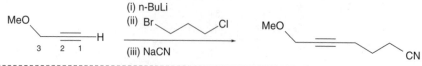

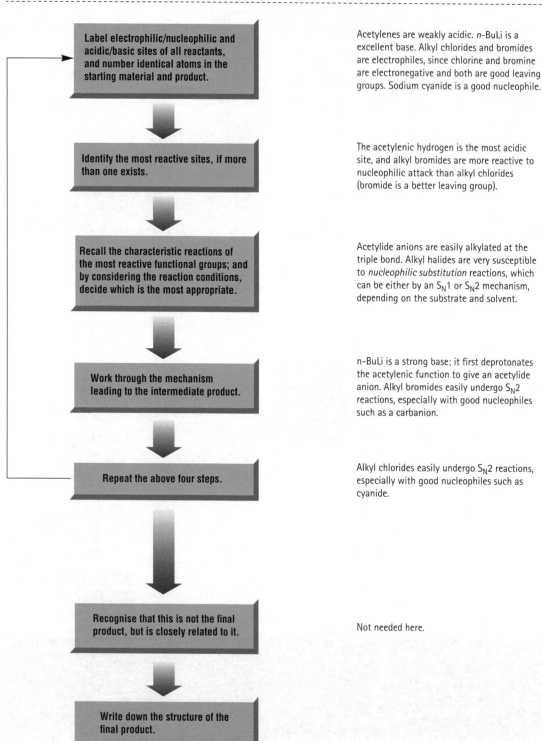

Label electrophilic/nucleophilic and acidic/basic sites of all reactants, and number identical atoms in the starting material and product.	Acetylenes are weakly acidic. *n*–BuLi is a excellent base. Alkyl chlorides and bromides are electrophiles, since chlorine and bromine are electronegative and both are good leaving groups. Sodium cyanide is a good nucleophile.
Identify the most reactive sites, if more than one exists.	The acetylenic hydrogen is the most acidic site, and alkyl bromides are more reactive to nucleophilic attack than alkyl chlorides (bromide is a better leaving group).
Recall the characteristic reactions of the most reactive functional groups; and by considering the reaction conditions, decide which is the most appropriate.	Acetylide anions are easily alkylated at the triple bond. Alkyl halides are very susceptible to *nucleophilic substitution* reactions, which can be either by an S_N1 or S_N2 mechanism, depending on the substrate and solvent.
Work through the mechanism leading to the intermediate product.	n–BuLi is a strong base; it first deprotonates the acetylenic function to give an acetylide anion. Alkyl bromides easily undergo S_N2 reactions, especially with good nucleophiles such as a carbanion.
Repeat the above four steps.	Alkyl chlorides easily undergo S_N2 reactions, especially with good nucleophiles such as cyanide.
Recognise that this is not the final product, but is closely related to it.	Not needed here.
Write down the structure of the final product.	

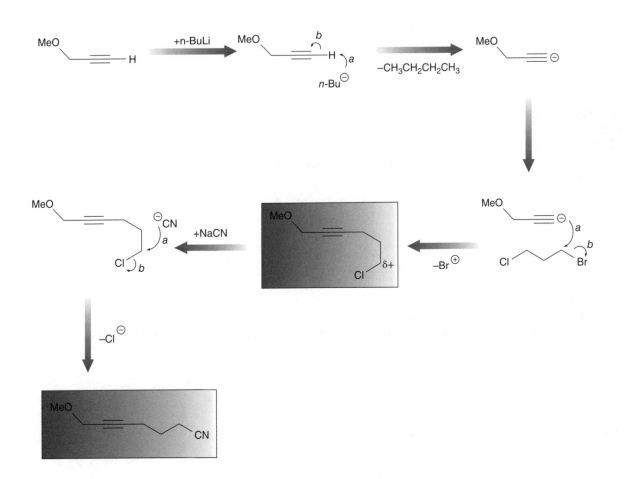

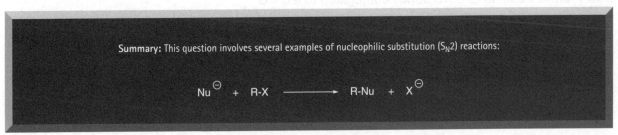

Summary: This question involves several examples of nucleophilic substitution (S_N2) reactions:

$$Nu^{\ominus} + R\text{-}X \longrightarrow R\text{-}Nu + X^{\ominus}$$

Now try questions 1.21 and 1.22

2 Alkene and alkyne chemistry

Alkene and alkyne bonding

The molecular orbital description of bonding in alkenes (sp^2 hybridised carbon) and alkynes (sp hybridised carbon).

Electrophilic addition

Addition of halogens

Mechanism and evidence for a stepwise process.
Nature of intermediates (cyclic bromonium and iodonium species).
Effect of substituents on reaction rate.
Stereochemistry of addition is *anti-*.

Addition of hydrogen halides

Mechanism and intermediacy of carbocations.
Orientation (Markovnikov's Rule) and the rationale for the effect of substitution on the orientation of addition.

Addition of hydrogen halides to conjugated dienes

1,2-versus 1,4-addition.

Nucleophilic addition to C=C bonds conjugated to a C=O

How to Solve Organic Reaction Mechanisms: A Stepwise Approach, First Edition. Mark G. Moloney.
© 2015 John Wiley & Sons, Ltd. Published 2015 by John Wiley & Sons, Ltd. Companion website: www.wiley.com/go/moloney/mechanisms

Other types of additions

Catalytic hydrogenation

Mechanism and stereochemistry: *cis*-addition of H_2.

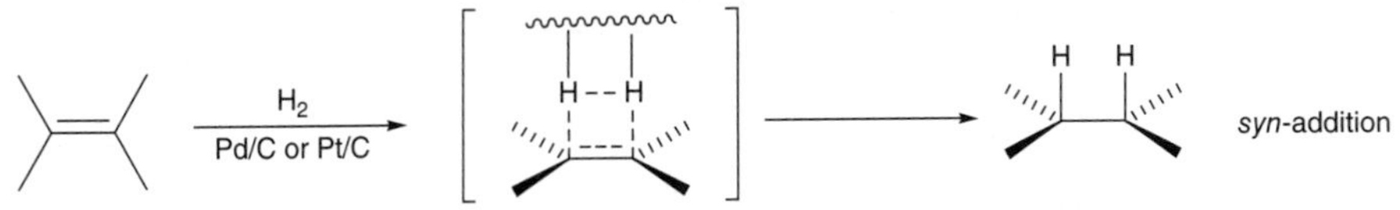

syn-addition

Dihydroxylation

Osmium tetroxide or alkaline permanganate gives *syn*-1,2-diols. These products can be cleaved to dicarbonyls by treatment with periodic acid or lead tetraacetate.

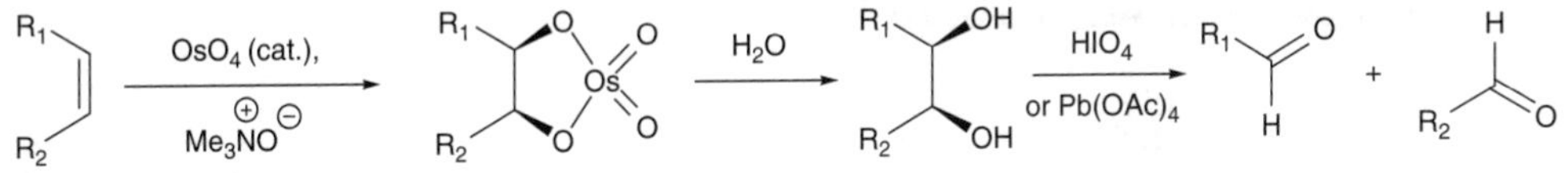

Epoxidation with peracids

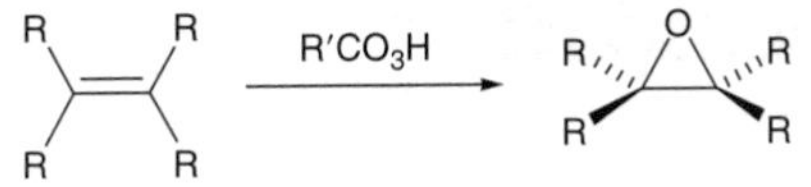

Ozonolysis with ozone

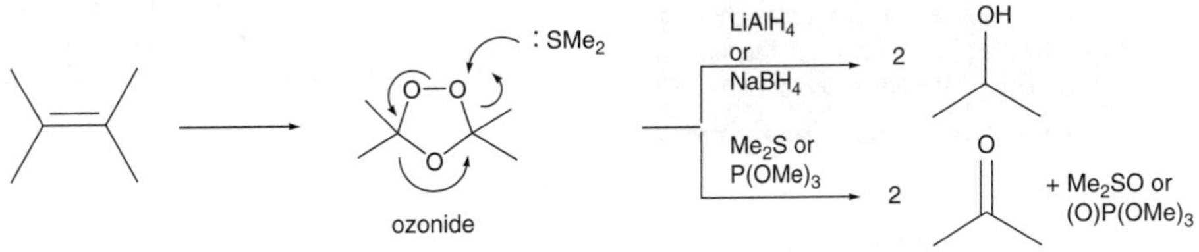

ozonide

Diels–Alder reaction

Concerted reaction of dienes and dienophiles (α,β-unsaturated carbonyl compounds).

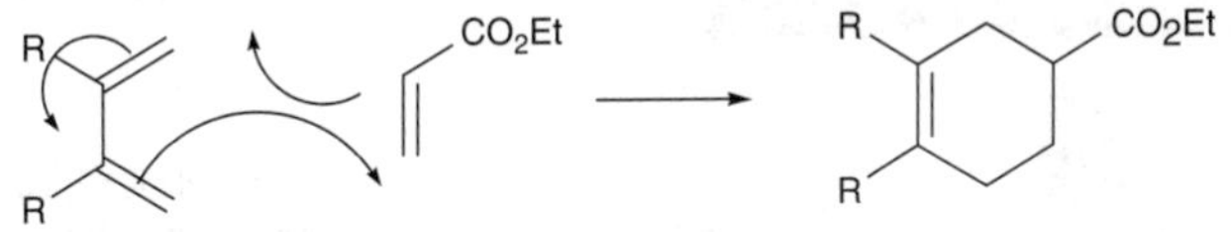

Hydroboration

Formation of a trialkylborane by *anti*-Markovnikov addition, and its conversion to an alcohol using alkaline hydrogen peroxide, and amines using hydroxylaminesulfonic acid.

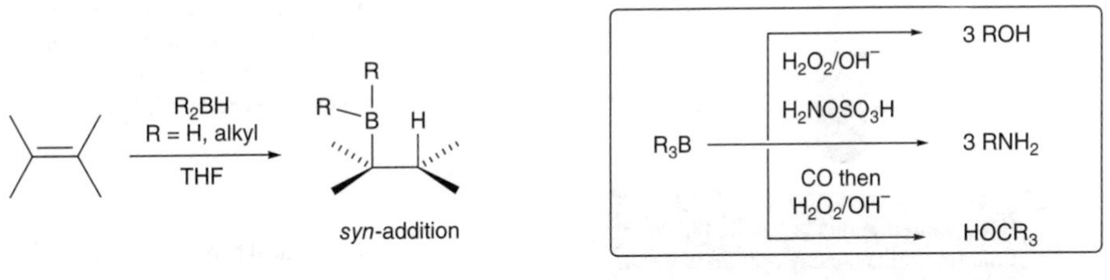

syn-addition

Reaction at the allylic position

Halogenation at the allylic position using *N*-bromosuccinimide and a radical initiator (e.g. $(PhCO_2)_2$).

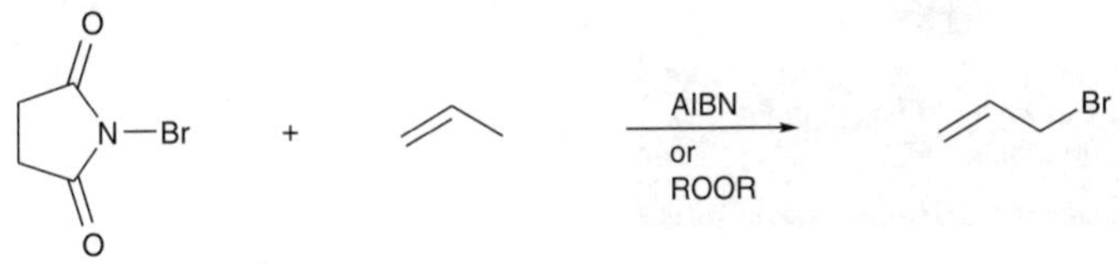

33

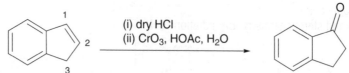

(i) dry HCl
(ii) CrO₃, HOAc, H₂O

| Label electrophilic/nucleophilic and acidic/basic sites of all reactants, and number identical atoms in the starting material and product. | Alkenes are good nucleophiles by virtue of their π electron density. HCl is a strong acid (pKₐ = −7). |

| Identify the most reactive sites, if more than one exists. | The cyclopentenyl double bond is the most reactive nucleophile, since the double bonds of the phenyl ring are stabilised by *aromaticity*. HCl is the only electrophile present. |

| Recall the characteristic reactions of the most reactive functional groups; and by considering the reaction conditions, decide which is the most appropriate. | Addition of HCl across the double bond occurs, with the orientation resulting from formation of the most stable carbocation (*Markovnikov's Rule*). |

| Work through the mechanism leading to the intermediate product. | Addition of a proton to C-2 of the alkene gives a stabilised *benzylic* carbocation, which is intercepted by chloride to give the product. |

| Repeat the above four steps. | * Chloride is a good leaving group, and can be displaced by the weak nucleophile, water, in an Sₙ1 process, favoured by the stabilised carbocation intermediate.

* The alcohol thus produced undergoes nucleophilic addition to chromium trioxide, to give a *chromate ester*, and in the process generates an excellent *leaving group* on oxygen. Collapse of the *chromate ester* occurs to give the ketone product. |

| Recognise that this is not the final product, but is closely related to it. | Not needed here. |

| Write down the structure of the final product. | |

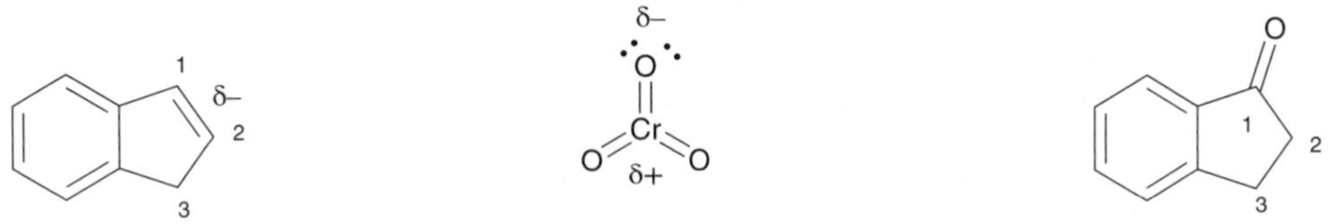
1
δ–
2
3
δ–
O
Cr
O
O
δ+
O
1
2
3

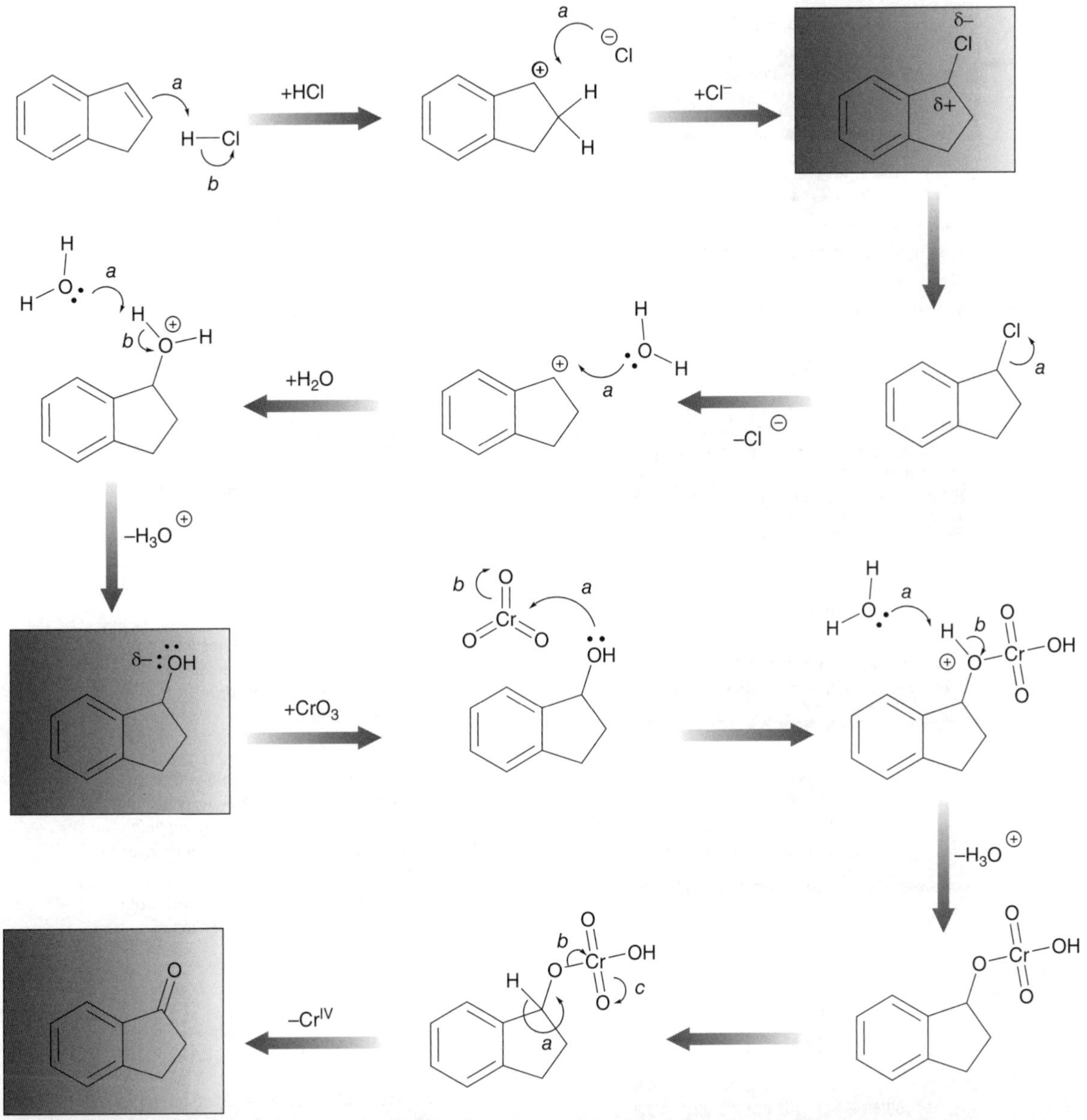
+HCl
+Cl⁻
+H2O
–Cl⊖
–H3O⊕
+CrO3
–H3O⊕
–Cr^IV

Summary: This is an example of Markovnikov addition to alkenes:

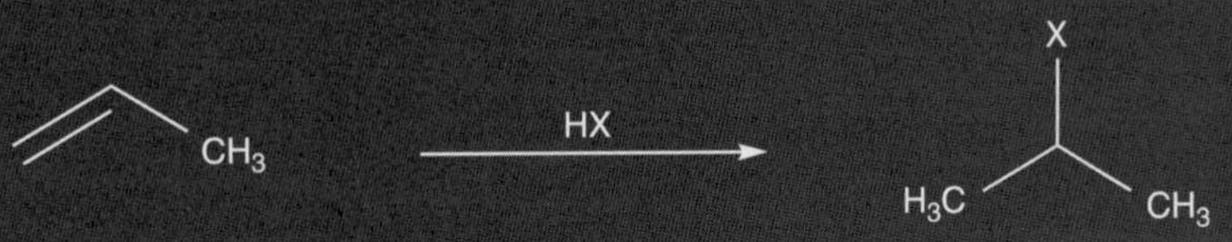
CH3
HX
X
H3C
CH3

Now try questions 2.8 and 2.9

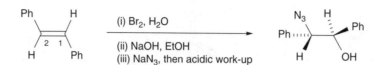

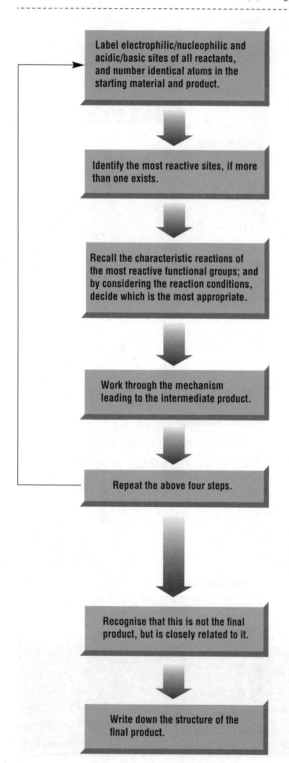

Bromine is an electrophile, with a weak Br-Br bond. Alkenes are nucleophiles by virtue of the π electron density.

The double bond is the most reactive nucleophile, since the double bonds of the phenyl rings are stabilised by *aromaticity*. Bromine is the only electrophile present. Water is a weak nucleophile.

Electrophilic addition of Br_2 to an alkene is easily possible; but in the presence of water, this leads to formation of a bromohydrin.

Addition of Br_2 gives an intermediate *bromonium* ion, which is intercepted by water to give a *bromohydrin*; the overall addition is *anti-*.

* Treatment of the bromohydrin with base gives the corresponding alkoxide, which undergoes an intramolecular *nucleophilic substitution*, to give an epoxide, with inversion of stereochemistry at C–1.
* *Nucleophilic substitution* (S_N2) with azide at the epoxide occurs with inversion of stereochemistry.

Protonation of the alkoxide anion gives the product, which needs to be re-drawn to show its identity with the original product structure.

Summary: This is an example of a diastereoselective anti-addition of an electrophile to an alkene:

Now try questions 2.10 and 2.16

2.3

 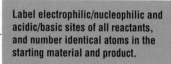

(i) dil H_2SO_4, Δ'
(ii) $PhCO_3H$

(iii) CF_3CO_2H, H_2O_2
(iv) KOH, H_2O

$Ph \overset{O}{\underset{}{\big\Vert}} Ph$ + $H_2C{=}O$

Label electrophilic/nucleophilic and acidic/basic sites of all reactants, and number identical atoms in the starting material and product.	Alcohols are nucleophiles and bases since the oxygen possesses lone pairs. Sulfuric acid is a strong acid, and fully ionised ($pK_a = -9$).
Identify the most reactive sites, if more than one exists.	Under strongly acidic conditions, the alcohol is the most basic site (although it is possible for an aromatic ring to be protonated, it is less basic since the double bonds of the phenyl rings are stabilised by *aromaticity*).
Recall the characteristic reactions of the most reactive functional groups; and by considering the reaction conditions, decide which is the most appropriate.	Alcohols readily eliminate water to give alkenes under acidic conditions.
Work through the mechanism leading to the intermediate product.	Protonation of the alcohol gives an *oxonium* ion, which is an excellent *leaving group*, and *elimination* to the alkene occurs; in this case, the reaction is E_1, proceeding through a stabilised *benzylic* carbocation.
Repeat the above four steps.	* Perbenzoic acid is a source of electrophilic oxygen, and easily *epoxidises* the alkene double bond.
	* Trifluoroacetic acid protonates the epoxide, which then opens (via the more stabilised *benzylic* carbocation in an S_N1 process); this cation is intercepted by hydrogen peroxide.
	* Hydroxide deprotonates the alcohol group, and induces fragmentation of the alkoxide gives the products, via cleavage of the weak O–O bond.
Recognise that this is not the final product, but is closely related to it.	Not needed here.
Write down the structure of the final product.	

Summary: This is an example of elimination and addition reactions of alkenes, and nucleophilic substitution reactions

Now try questions 2.11 and 2.17

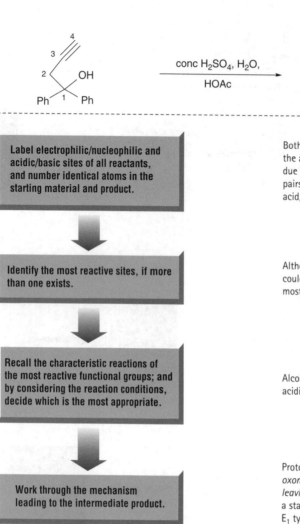

Label electrophilic/nucleophilic and acidic/basic sites of all reactants, and number identical atoms in the starting material and product.	Both the phenyl and alkyne groups, and the alcohol, are nucleophilic and basic, due to the π electron density and lone pairs respectively. Sulfuric acid is a strong acid, and fully ionised ($pK_a = -9$).
Identify the most reactive sites, if more than one exists.	Although the phenyl ring and the alkyne could also be protonated, the alcohol is the most basic site.
Recall the characteristic reactions of the most reactive functional groups; and by considering the reaction conditions, decide which is the most appropriate.	Alcohols readily eliminate water under acidic conditions to give alkenes.
Work through the mechanism leading to the intermediate product.	Protonation of the alcohol to give an *oxonium* cation occurs; this is a good *leaving group*, and departure generates a stabilised *benzylic* carbocation in an E_1 type process. *Elimination* then occurs to the alkene product.
Repeat the above four steps.	Alkynes undergo *electrophilic addition* reactions, and under aqueous acidic conditions are hydrolysed to the ketone product. Protonation of the alkyne at the terminal position gives the more substituted *vinylic* cation, which is intercepted by water. The resulting enol *tautomerises* to the ketone product.
Recognise that this is not the final product, but is closely related to it.	Not needed here.
Write down the structure of the final product.	

Summary: This is an example of the acid-catalysed hydration of an alkyne:

Now try questions 2.12 and 2.18

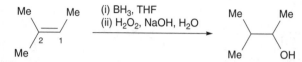

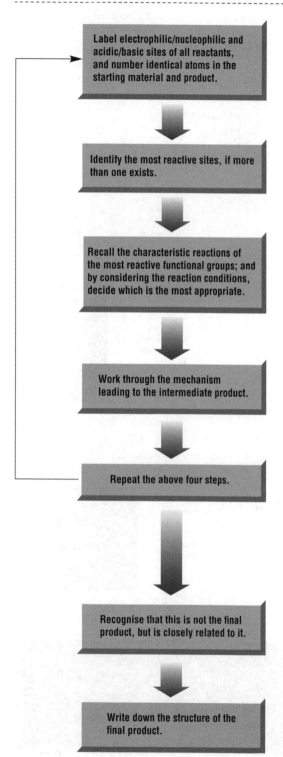

Alkenes are nucleophiles by virtue of the π electron density. Borane is an electrophile (boron is electron deficient) and is therefore a good Lewis acid; it forms stable B–O bonds.

The alkene is the only reactive nucleophilic site, and boron is the only electrophile.

Hydroboration of alkenes occurs easily, involving addition of B–H across the double bond, with boron adding to the less hindered carbon.

The *addition* of H and B to the alkene occurs in a *syn*-process; this is repeated a further two times using the remaining B–H bonds, to give a *trialkylborane*.

The boron atom of the trialkylborane is still electron deficient, and is readily attacked by hydroperoxide anion (generated from hydrogen peroxide and base). This process induces a *1,2–alkyl migration* from to B to O, which is repeated a further two times to give a *trialkoxyborane*.

The *trialkoxyborane* is easily *hydrolysed* under alkaline conditions to the corresponding alkoxide, which is protonated under the aqueous conditions of the reaction.

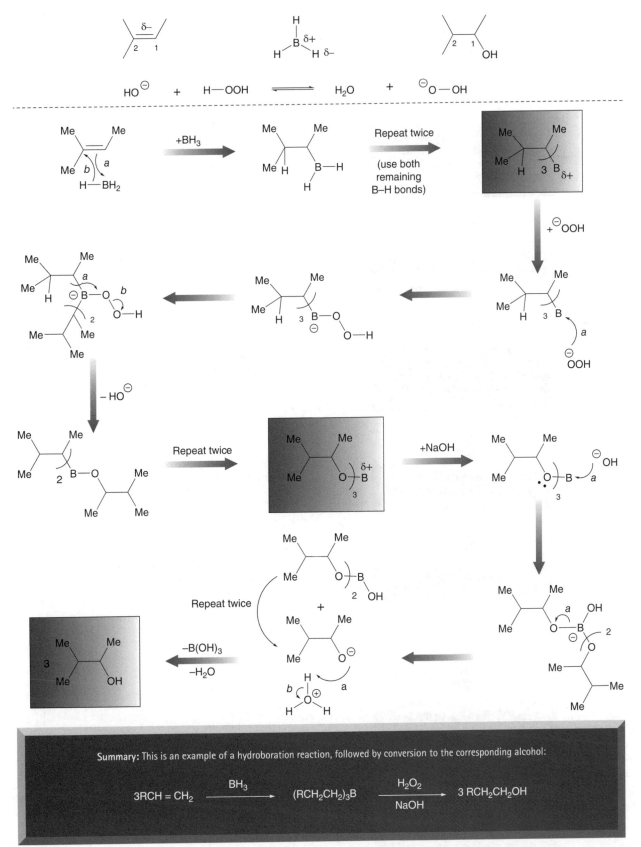

Summary: This is an example of a hydroboration reaction, followed by conversion to the corresponding alcohol:

$$3RCH = CH_2 \xrightarrow{\text{BH}_3} (RCH_2CH_2)_3B \xrightarrow[\text{NaOH}]{\text{H}_2\text{O}_2} 3\ RCH_2CH_2OH$$

Now try questions 2.13 and 2.19

2.6

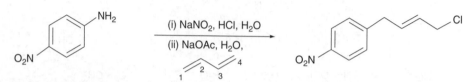

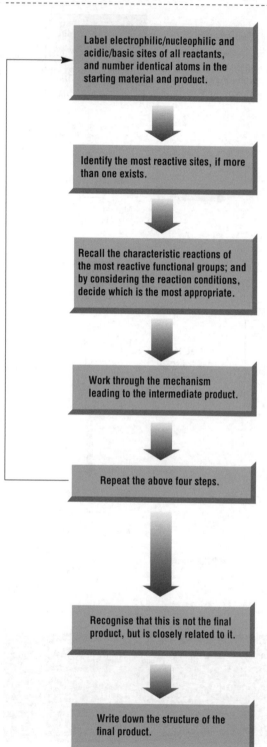

The amine group is a good nucleophile, although in this case, not; it is deactivated by *resonance* with the nitro group on the aromatic ring. Under acidic conditions, sodium nitrite effectively generates the nitrosonium cation ($O=N^+$), a good electrophile.

Although the phenyl ring is nucleophilic, the amine is the most nucleophilic site. The nitrosonium cation is the most reactive electrophile.

The combination $NaNO_2$/HCl generates HONO (nitrous acid), which is in *equilibrium* with N_2O_3, effectively a source of $O=N^+$. This reagent in turn *diazotises* the amine functional group.

Nucleophilic addition of the amine to N_2O_3 is followed by tautomerisation, which permits *elimination* of H_2O to give the *diazonium* salt product.

The diazo group loses nitrogen gas to give an aryl cation; this is intercepted by the diene at the chain terminus to give a resonance stabilised *allylic* cation. The 2° *allylic* intermediate cation shown is the more stable. The *allylic* cation reacts with chloride at the least hindered end to give the product of 1,4-addition.

Not needed here.

$$ArN_2^{\oplus} \xrightarrow{\quad Nu^{\ominus} \quad} ArNu$$

Now try questions 2.14 and 2.20

2.7

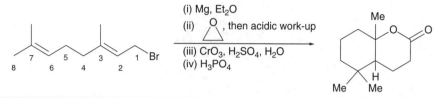

(i) Mg, Et$_2$O
(ii) [epoxide] , then acidic work-up
(iii) CrO$_3$, H$_2$SO$_4$, H$_2$O
(iv) H$_3$PO$_4$

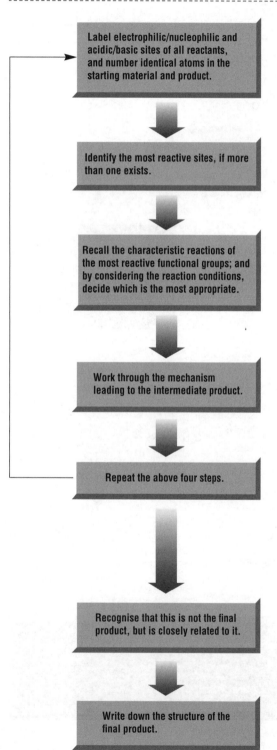

Alkyl halides and magnesium react to form Grignard reagents, which are good nucleophiles. Ethylene oxide is an electrophile, with the C–O bond polarised by the electronegative oxygen; there is substantial ring strain present.

The alkyl halide reacts at C-1 to generate the corresponding Grignard, a carbanion equivalent, which is a good nucleophile; this is the only nucleophilic site. Ethylene oxide is susceptible to nucleophilic attack at the carbon adjacent to the oxygen; this is the only electrophilic site.

Grignard reagents react with ethylene oxide to give alcohols in which the alkyl chain has been extended by 2 carbon atoms.

Nucleophilic attack (S$_N$2) attack by the Grignard at ethylene oxide gives an alkoxide intermediate; protonation on acid work-up gives an alcohol.

* The alcohol undergoes nucleophilic addition to chromium trioxide, to give a chromate ester which collapses in an elimination reaction to the aldehyde; this aldehyde is in *equilibrium* with its *hydrate*, formed by the addition of water the carboonyl group, and the oxidation process with chromium trioxide is repeated, to give a carboxylic acid.
* Phosphoric acid, a strong acid, protonates the most electron rich terminal double bond, giving a stabilised tertiary carbocation (*Markovnikov addition*). The carboxylic acid function acts as an internal nucleophile, to give the bicyclic lactone product.

Not needed here.

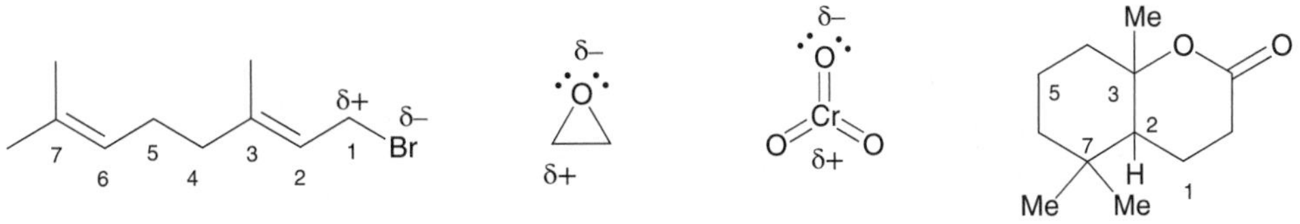

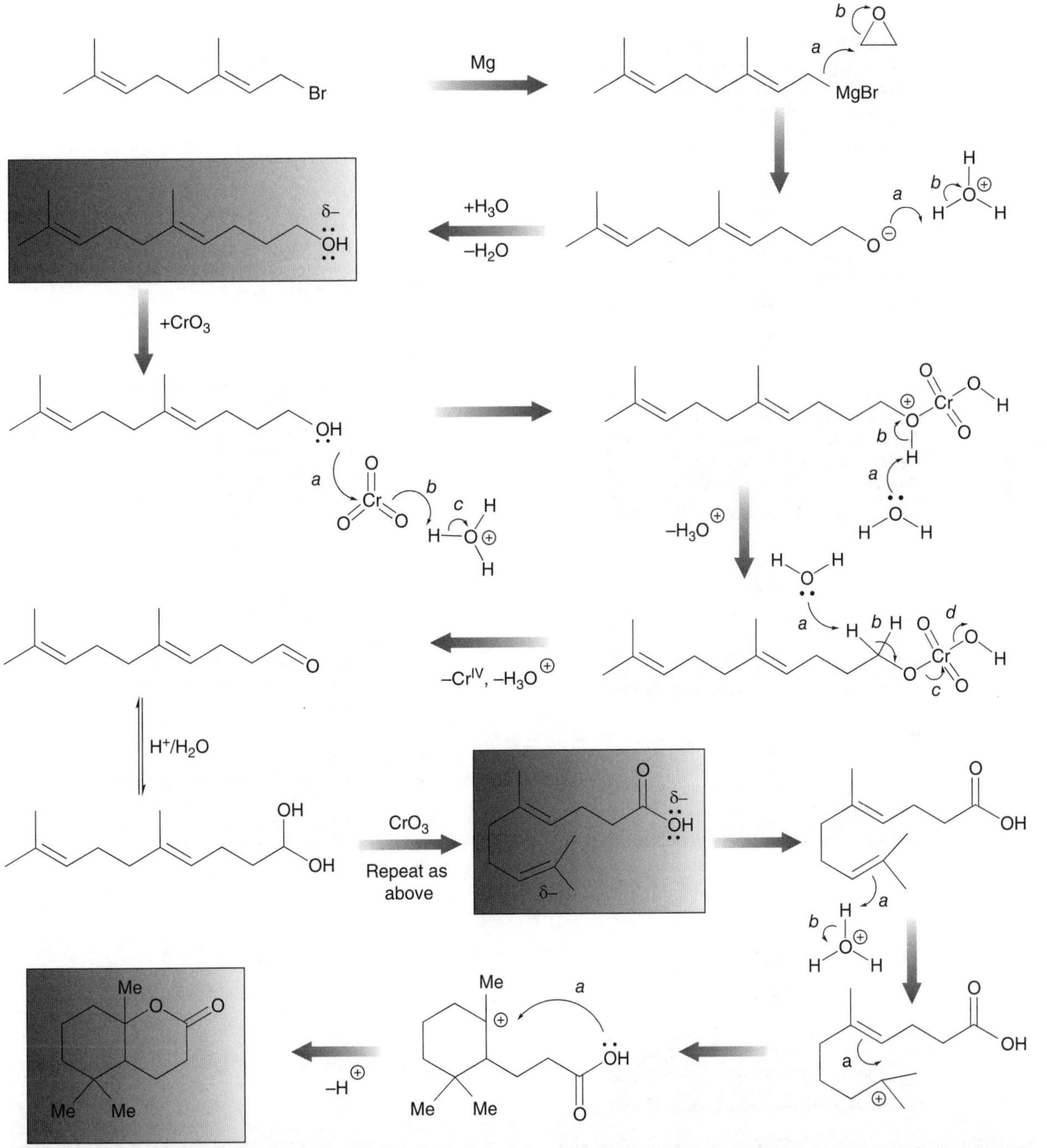

$\delta-$

$\delta+$

$\delta+$ $\delta-$

$\delta-$

$\delta+$

7 6 5 4 3 2 1 Br

Me

5 3 2

7 H

Me Me 1

Mg

b O

a

MgBr

+H$_3$O

−H$_2$O

$\delta-$ OH

a

b H

H O H

O

+CrO$_3$

OH

a Cr b c H

O O H O H

O Cr O H

b

a H

H O H

O O

−H$_3$O$^\oplus$

H O H

a H b H O d Cr O H

O

c

O

−Cr^{IV}, −H$_3$O$^\oplus$

H$^+$/H$_2$O

OH

OH

CrO$_3$

Repeat as above

O

$\delta-$ OH

$\delta-$

O

OH

b H

H O H

O

a

O

OH

a

$\oplus$

Me

$\oplus$

a

Me Me

OH

O

Me

O O

Me Me

−H$^\oplus$

Summary: This is an example of the intramolecular trapping of an alkyl cation with a carboxylic acid to give a lactone.

Now try questions 2.15, 2.21 and 2.22

47

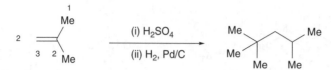

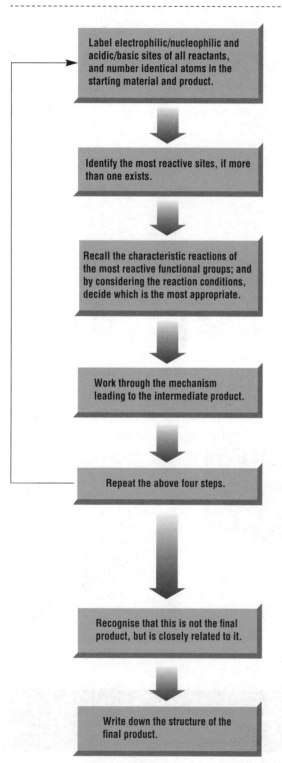

Alkenes are good nucleophiles by virtue of their π electron density. Sulfuric acid is a strong acid, and fully ionised (pK$_a$ = −9).

The double bond is the only nucleophilic site and the proton released from the sulfuric acid is the only electrophile.

Alkenes are readily protonated by strong acids to give *carbocations*, especially if they are stabilised by electron releasing groups, in this case 2 methyl groups; the *regioselectivity* of the protonation is such that the most stable tertiary carbocation is formed, instead of the alternative primary carbocation.

Protonation of the double bond leads to the formation of a tertiary carbocation. This electrophile is in turn intercepted by another molecule of starting alkene, itself a nucleophile, which generates another tertiary carbocation. *Elimination* gives the product alkene.

Alkenes are good nucleophiles by virtue of their π electron density, but may be readily *reduced* under a hydrogen atmosphere in the presence of a palladium-charcoal catalyst, to give alkanes arising by stereospecific *syn*-addition of hydrogen.

Not needed here.

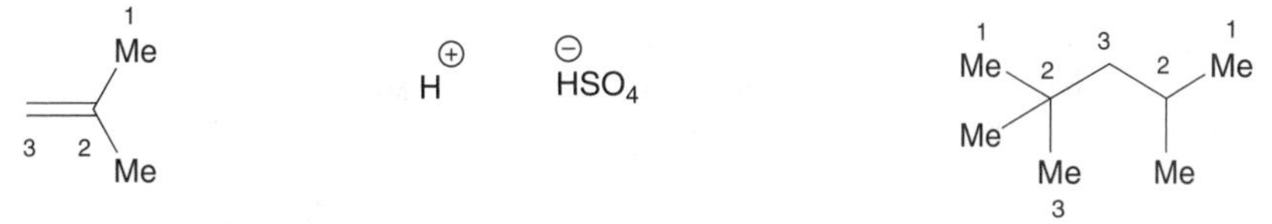

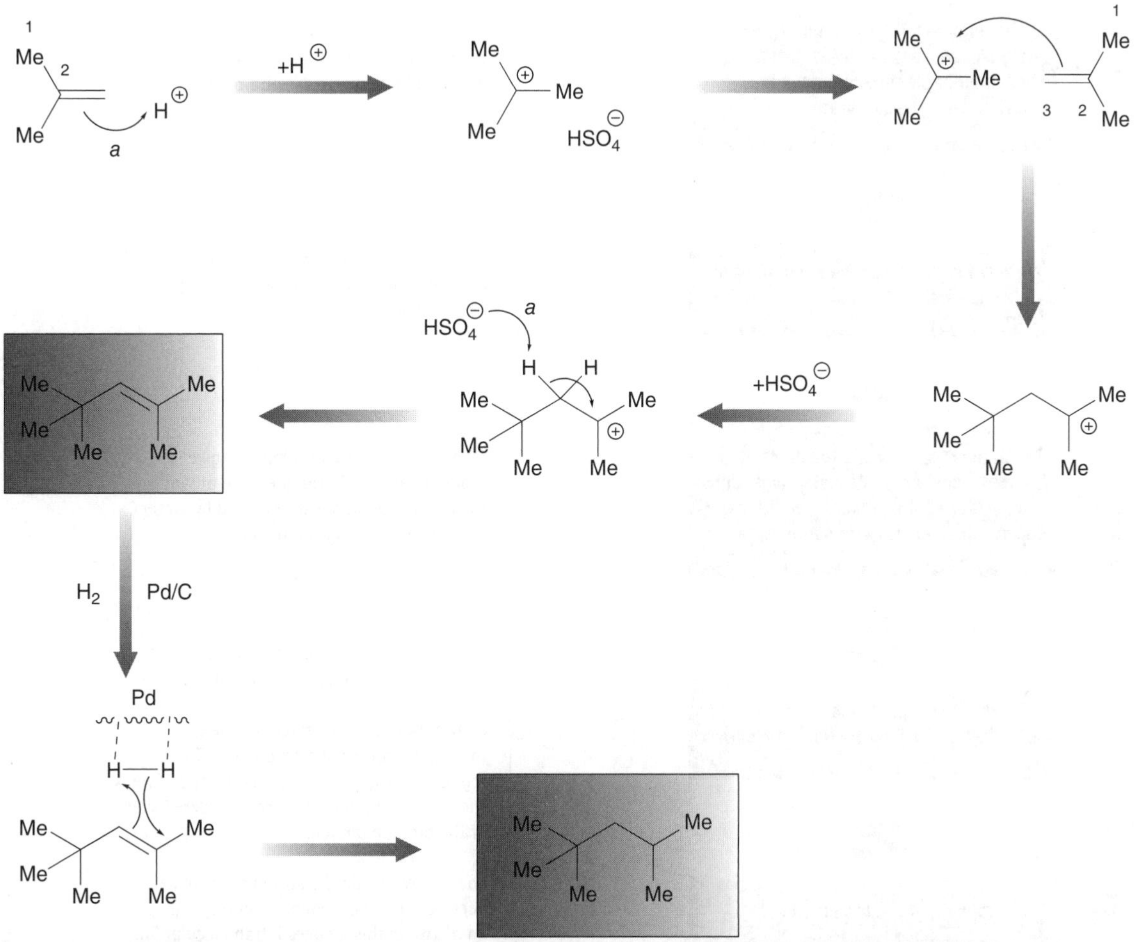

Me

1

Me
3   2
Me

H $\oplus$ $\ominus$ HSO$_4$

1   3   1
Me 2   Me 2   Me
Me   Me   Me
            3

1
Me 2
Me

a H $\oplus$

+H $\oplus$

Me
$\oplus$ Me
Me $\ominus$ HSO$_4$

Me
$\oplus$ Me
Me   3   2
            Me

Me
Me
Me   Me

$\oplus$

+HSO$_4$ $\ominus$

Me   Me
Me $\oplus$
Me   Me

HSO$_4$ $\ominus$ a

H   H
Me   Me
Me $\oplus$
Me   Me

Me   Me
Me   Me   Me

H$_2$ Pd/C

Pd

H—H

Me   Me
Me   Me   Me

Me   Me
Me   Me
Me   Me

Summary: This is an example of Markovnikov addition to alkenes:

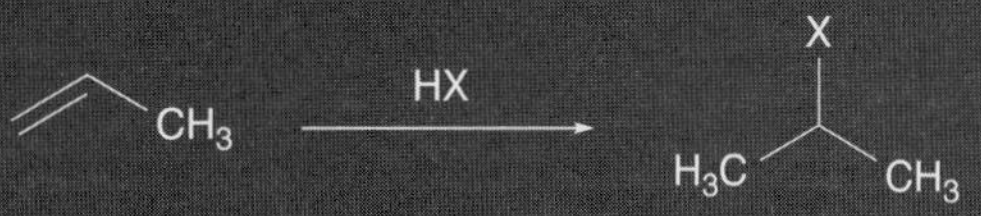

CH$_3$ HX   →   X
H$_3$C   CH$_3$

Now try question 2.9

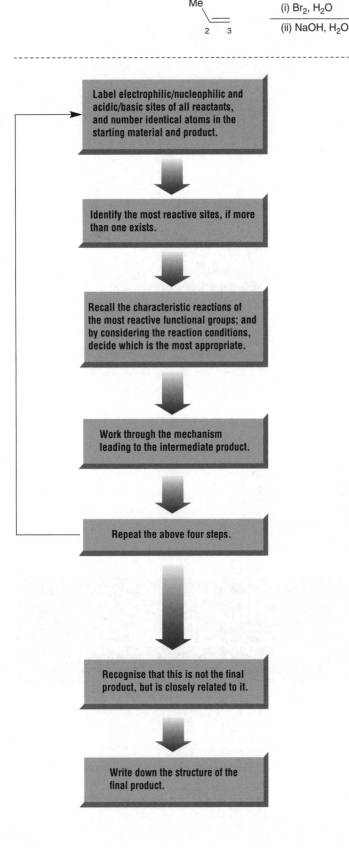

Alkenes are good nucleophiles by virtue of their π electron density. Bromine is a weak electrophile and water is also a weak nucleophile.

The double bond and water are the only nucleoophilics and bromine is the only electrophile.

Alkenes readily react bromine to give bromonium ions; in the presence of water, the bromonium ion is intercepted by water to give a bromohydrin product.

Bromination of the alkene leads to the formation of a bromonium ion; this species opens to give the more stabilised 2° carbocation. This is intercepted by water, to give the corresponding product. The *regioselectivity* of the protonation therefore is such that water adds the most hindered side of the bromonium ion.

Sodium hydroxide is a strong base, and deprotonates the alcohol function to give an *alkoxide*; this induces intramolecular S_N2 attack of the alkoxide nucleophile with displacement of the bromide leaving group, leading to cyclisation.

Not needed here.

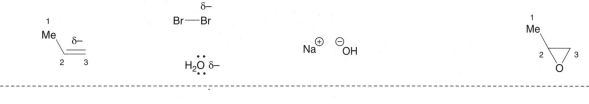

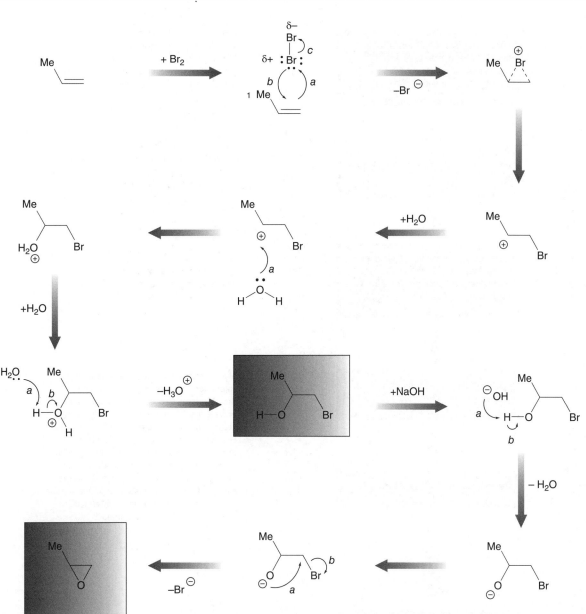

Summary: This is an example of nucleophilic substitution reactions and addition reactions of alkenes

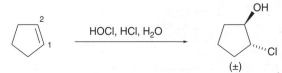

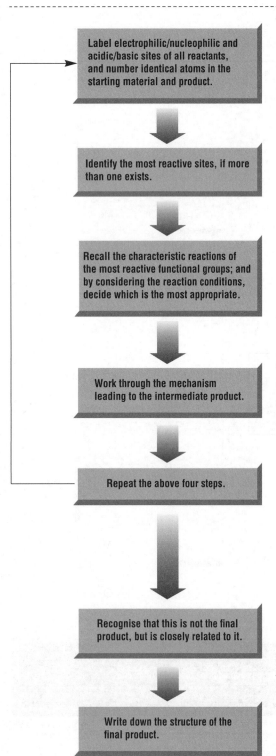

| Label electrophilic/nucleophilic and acidic/basic sites of all reactants, and number identical atoms in the starting material and product. | Alkenes are good nucleophiles by virtue of their π electron density. HCl is a strong acid and fully ionised (pK$_a$ = −7). This establises an equilibrium with hypochlorous acid (HOCl) which provides a source of chlorine. |

Identify the most reactive sites, if more than one exists.

The carbon-carbon double bond and water are both nucleophiles and chlorine is the only electrophile.

Recall the characteristic reactions of the most reactive functional groups; and by considering the reaction conditions, decide which is the most appropriate.

Chlorine water adds to alkenes to give chlorohydrin products.

Work through the mechanism leading to the intermediate product.

Repeat the above four steps.

Alkenes readily react with chlorine to give chloronium ions, which prefer to exist in the *open* form, that is, with a carbocation intermediate; initial addition of chlorine may occur at the top or bottom face, which gives enantiomeric intermediate carbocations. Either of these intermediates may be intercepted by water to give a chlorohydrin product, with an *anti*-outcome in which water attacks at the opposite side to chlorine, and as a racemic pair. Deprotonation of the oxonium intermediate generates the chlorohydrin product.

Recognise that this is not the final product, but is closely related to it.

Not needed here.

Write down the structure of the final product.

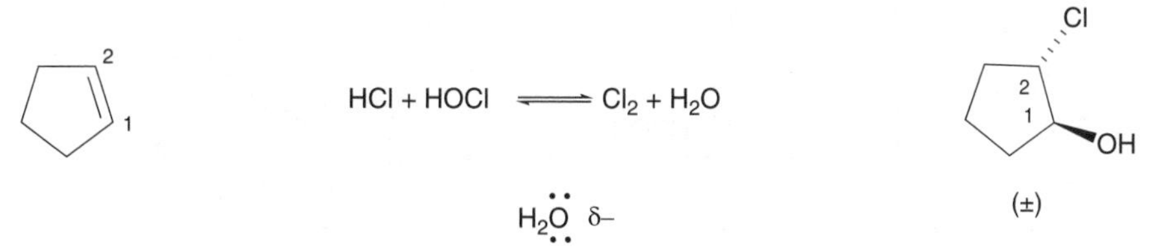

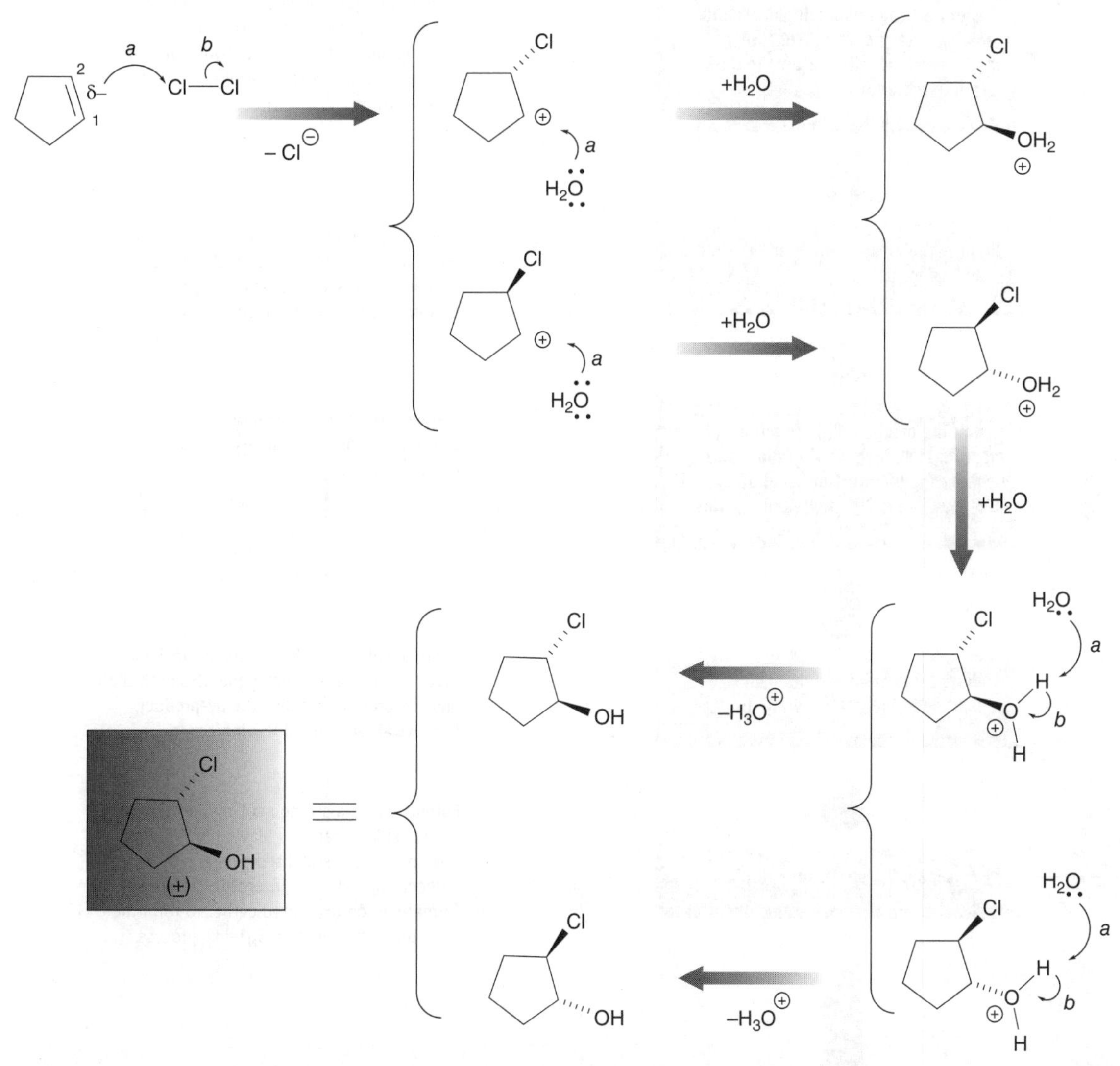

HCl + HOCl $\rightleftharpoons$ Cl$_2$ + H$_2$O

$H_2\overset{..}{\underset{..}{O}}$ $\delta-$

(±)

a b

$\delta-$

$-$ Cl$^{\ominus}$

+H$_2$O

+H$_2$O

+H$_2$O

$-$H$_3$O$^{\oplus}$

$-$H$_3$O$^{\oplus}$

(±)

$\equiv$

Summary: This is an example of nucleophilic substitution reactions, and elimination and addition reactions of alkenes

Now try question 2.16

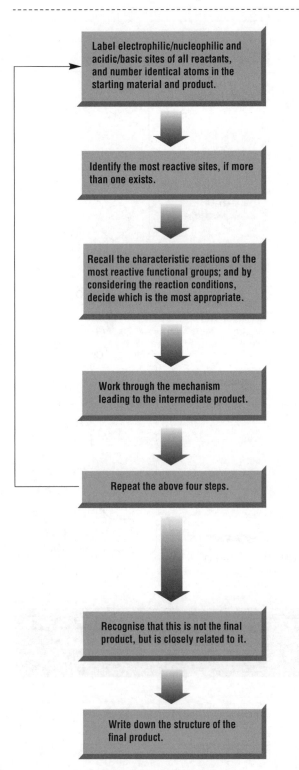

Label electrophilic/nucleophilic and acidic/basic sites of all reactants, and number identical atoms in the starting material and product.	Alkenes are good nucleophiles by virtue of their π electron density. Performic acid is electrophilic at the terminal hydroxyl group, as a result of the weak O–O bond and the good leaving group.
Identify the most reactive sites, if more than one exists.	The cyclopentenyl double bond is the most reactive, since the double bonds of the phenyl ring are stabilised by *aromaticity*, and performic acid the only electrophile.
Recall the characteristic reactions of the most reactive functional groups; and by considering the reaction conditions, decide which is the most appropriate.	Alkenes are readily epoxidised by electrophilic oxygen sources such as performic acid.
Work through the mechanism leading to the intermediate product.	Oxygen delivery to the alkene occurs by a cyclic mechanism, which generates both the epoxide product but also the by-product, formate anion, a good leaving group.
Repeat the above four steps.	Formic acid is a strong acid, and protonates the epoxide, creating a leaving group. Ring opening gives a stabilised benzylic carbocation, which is intercepted by the formate anion generated earlier to form the 1, 2- product; this is an S_N1-like process.
Recognise that this is not the final product, but is closely related to it.	Not needed here.
Write down the structure of the final product.	

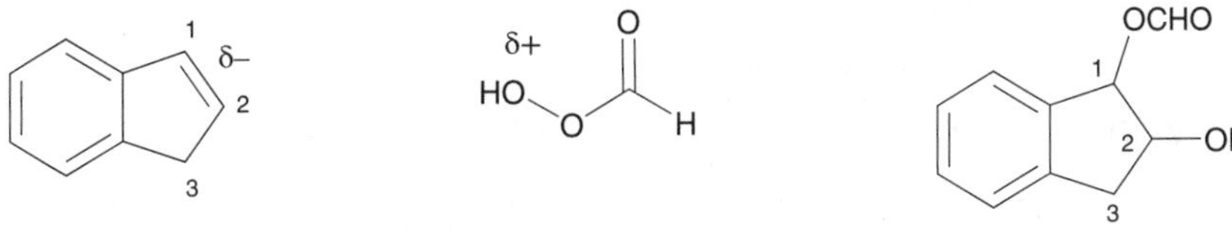

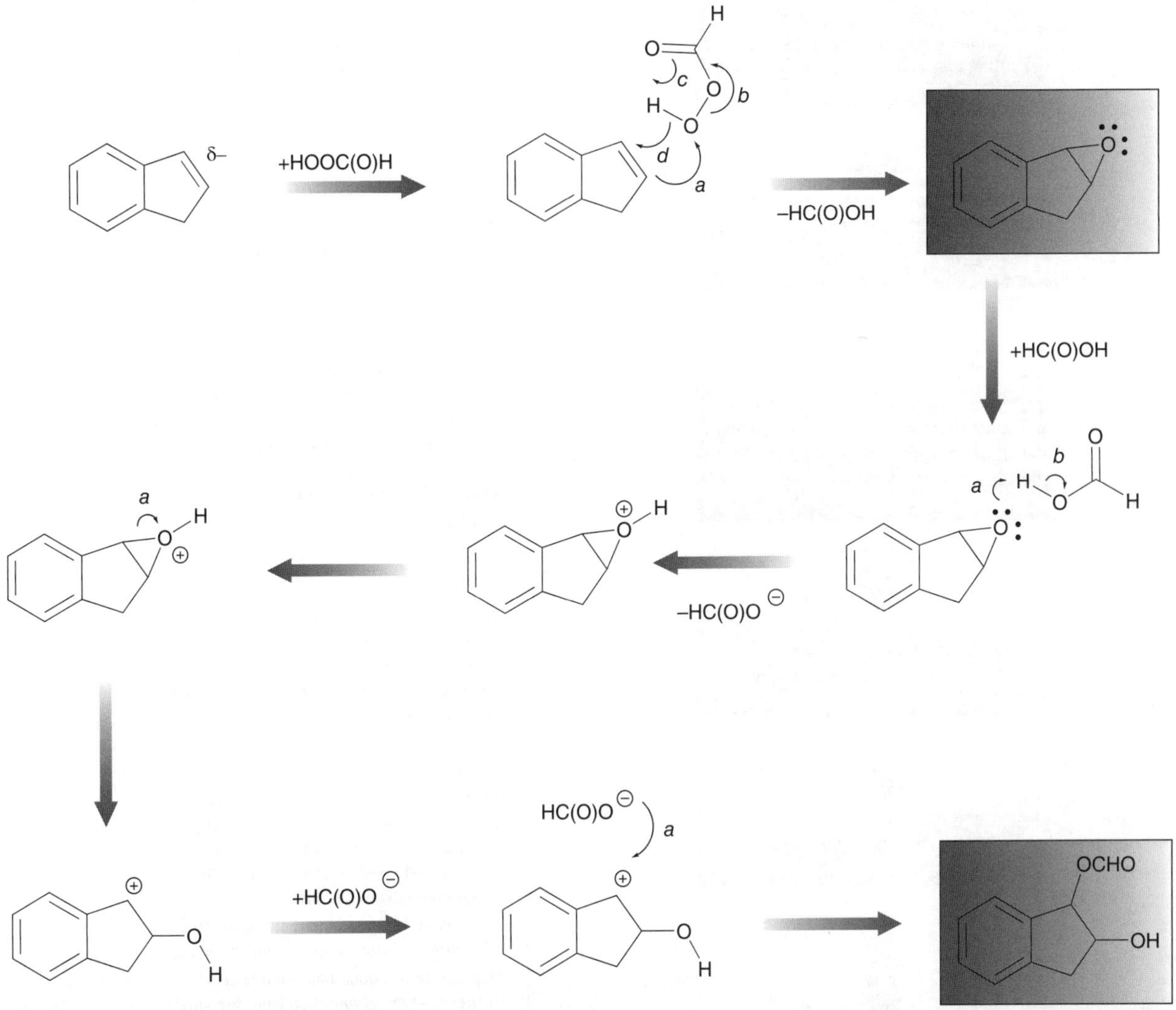

$\delta-$

$\delta+$

OCHO

1
2
3

HO—O

1
2
3

—OH

$\delta-$

$+HOOC(O)H$

$-HC(O)OH$

$+HC(O)OH$

$-HC(O)O^{\ominus}$

$+HC(O)O^{\ominus}$

$HC(O)O^{\ominus}$

OCHO

—OH

Summary: This is an example of epoxidation and ring opening reactions:

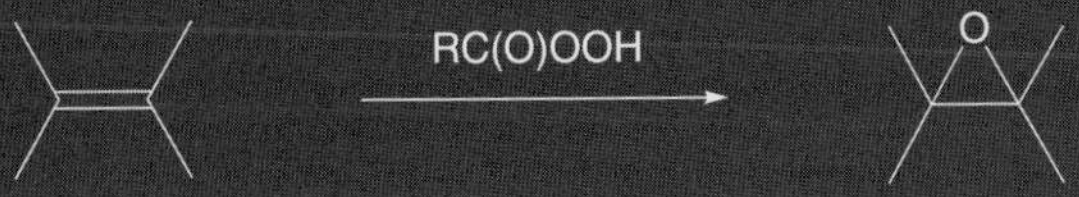

$RC(O)OOH$

Now try question 2.17

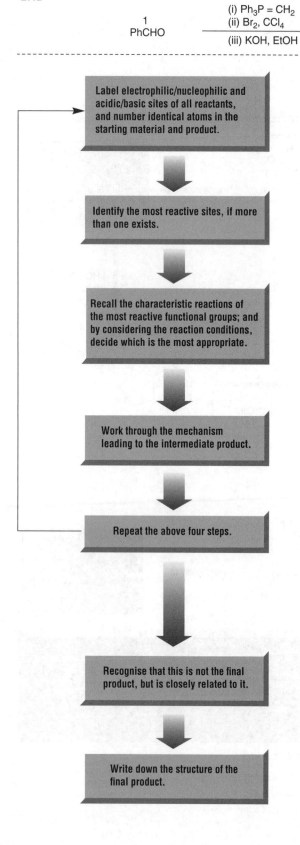

$$\underset{\text{PhCHO}}{\overset{1}{}} \xrightarrow[\text{(iii) KOH, EtOH}]{\overset{\text{(i) Ph}_3\text{P} = \text{CH}_2}{\text{(ii) Br}_2, \text{CCl}_4}} \text{Ph} \equiv\!\!\!=\!\!\!= \text{H}$$

Label electrophilic/nucleophilic and acidic/basic sites of all reactants, and number identical atoms in the starting material and product.

Identify the most reactive sites, if more than one exists.

Recall the characteristic reactions of the most reactive functional groups; and by considering the reaction conditions, decide which is the most appropriate.

Work through the mechanism leading to the intermediate product.

Repeat the above four steps.

Recognise that this is not the final product, but is closely related to it.

Write down the structure of the final product.

Aldehydes are electrophiles, since oxygen is more electronegative than carbon. *Ylides* are good nucleophiles, being polarised with a negative charge on carbon.

The aldehyde is the only electrophilic site and the ylide is the only nucleophilic one.

Aldehydes are very susceptible to *nucleophilic addition* reactions, and especially by carbon nucleophiles. Phosphonium ylides participate in the Wittig reaction, converting an aldehyde into an alkene, in a process which is driven by formation of the strong P=O bond.

Nucleophilic addition by the ylid to the aldehyde gives a cyclic oxaphosphetane intermediate; this collapses with loss of the very stable triphenylphosphine oxide to give the alkene product.

* Bromine is an electrophile; alkenes are good nucleophiles by virtue of their π electron density. The two react together via formation of a bromonium ion which is intercepted by the resulting bromide anion to give the 1,2-dibromoalkane product.
* Hydroxide is a good base, and leads to E2-type *elimination* of HBr to give a *vinyl bromide*.
*Hydroxide is a good base, and leads to further E2-type *elimination* from the vinyl bromide of HBr to give an alkyne.

Not needed here.

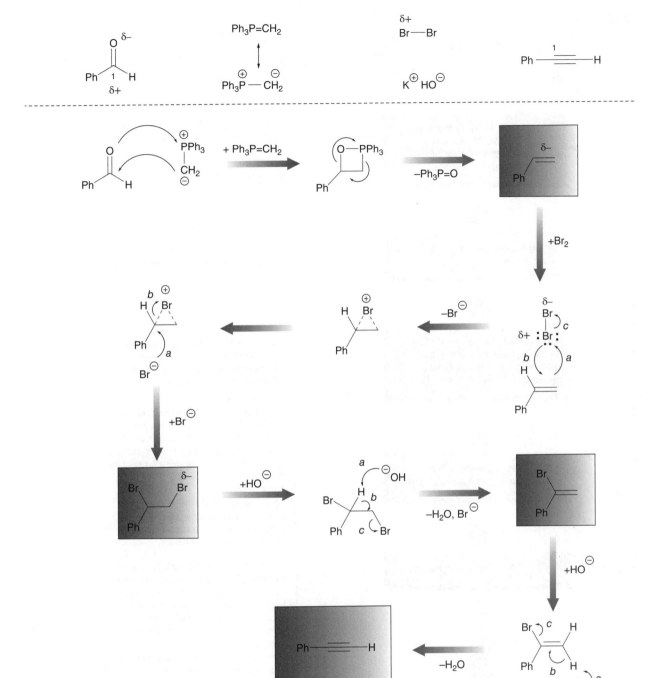

Summary: This is an example of addition and elimination reactions of alkenes

Now try question 2.18

2.13

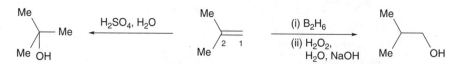

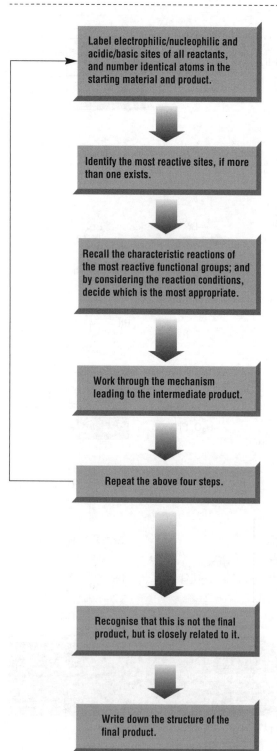

Label electrophilic/nucleophilic and acidic/basic sites of all reactants, and number identical atoms in the starting material and product.	Alkenes are nucleophiles by virtue of the π electron density. Borane is an electrophile (boron is electron deficient) and is therefore a Lewis acid. Sulfuric acid is a strong acid, and fully ionised ($pK_a = -9$).
Identify the most reactive sites, if more than one exists.	The alkene is the only reactive nucleophile, and boron and acid the only electrophiles.
Recall the characteristic reactions of the most reactive functional groups; and by considering the reaction conditions, decide which is the most appropriate.	Alkenes are very susceptible to *electrophilic addition* reactions, which with appropriate sources of oxygen lead to epoxidation and 1,2-dihydroxylation products.
Work through the mechanism leading to the intermediate product.	* *Hydroboration* of alkenes occurs easily (right hand arrow), adding B–H across the carbon–carbon double bond, with boron adding to the less hindered carbon and in a *syn*- process; this is repeated a further two times using the remaining B–H bonds, to give a *trialkylborane*. * A proton adds to the alkene (left hand arrow) in a Markovnikov sense, to give a carbocation which is intercepted by water to give a tertiary alcohol product.
Repeat the above four steps.	* The boron atom of the trialkylborane is still electron deficient, and is readily attacked by hydroperoxide anion (generated from hydrogen peroxide and base). This process induces a *1,2-alkyl migration* from to B to O, which is repeated a further two times to give a *trialkoxyborane*. * The *trialkoxyborane* is easily *hydrolysed* under alkaline conditions to the corresponding alkoxide, which is protonated under the acidic work-up (see 2.5).
Recognise that this is not the final product, but is closely related to it.	Not needed here
Write down the structure of the final product.	

Me

$H_3O^{\oplus}$ $HSO_4^{\ominus}$

Me 2 1

$H_2\ddot{O}$ $\delta-$

H B $\delta+$
H H $\delta-$

Me
   2 1
Me   OH

Me
   Me Me
   2 1
Me OH

$HO^{\ominus}$ + H—OOH $\rightleftharpoons$ H_2O + $^{\ominus}O$—OH

- -

Me
        + BH$_3$ → Me
Me   Me
                    b ↶ ↷ a
                    H—BH$_2$

Me
Me   BH$_2$

Repeat twice
(use both remaining B–H bonds)

Me
Me   3 BH$_2$ $\delta+$

+$H_3O^{\oplus}$

Me
Me

a b H
   O$^{\oplus}$
H   H

+ $^{\ominus}$OOH

Me
Me   3 BH$_2$

HOO$^{\ominus}$ a

Me
Me   3 B—O—OH
        $^{\ominus}$

Me
Me   2 B—O b
        a   OH
        $^{\ominus}$
   Me
   Me

a Me
   $^{\oplus}$ Me
$H_2\ddot{O}$ Me

+H_2O

−HO$^{\ominus}$

Me
Me   2 B—O Me
            Me

Repeat twice
(use both remaining B–C bonds)

Me
Me   O B
        $\ddot{}$ 3
           a $^{\ominus}$OH

Me
Me   O—BH a OH
        $^{\ominus}$ 2
           O
           Me

a H
   O$^{\oplus}$ b
$H_2\ddot{O}$ Me H
   Me Me
   Me

−H$_3$O$^+$ → Me
              Me
              Me OH

Repeat twice + $^{\ominus}$OH

Me
Me   3
Me   OH

+NaOH
−B(OH)$_3$

Me
Me   O B
           3

Summary: This is an example of a hydroboration reaction, followed by conversion to the corresponding alcohol:

$$3\ RCH = CH_2 \xrightarrow{BH_3} (RCH_2CH_2)_3B \xrightarrow[NaOH]{H_2O_2} 3\ RCH_2CH_2OH$$

Now try question 2.19

2.14

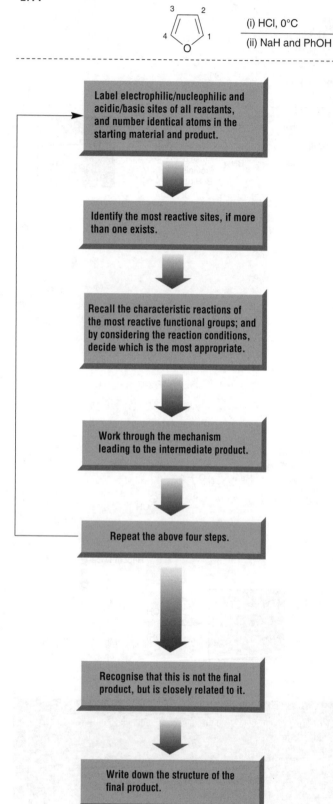

(i) HCl, 0°C

(ii) NaH and PhOH

Label electrophilic/nucleophilic and acidic/basic sites of all reactants, and number identical atoms in the starting material and product.

Furans a weakly aromatic, but are good nucleophiles by virtue of the resonance releasing oxygen substituent. HCl is a strong acid and fully ionised ($pK_a = -7$).

Identify the most reactive sites, if more than one exists.

The C-4 position is the most nucleophilic, since it is activated by resonance release from the oxygen of the furan ring.

Recall the characteristic reactions of the most reactive functional groups; and by considering the reaction conditions, decide which is the most appropriate.

Alkenes are nucleophiles by virtue of the π electron density. A proton is an electrophile and will readily protonate a double bond.

Work through the mechanism leading to the intermediate product.

The furan system is readily protonated at the C-4 position to give an *oxonium* ion intermediate. This is intercepted by chloride to give the α-chloroether product.

Repeat the above four steps.

Sodium hydride is a strong base; it readily deprotonates the alcohol function of phenol to generate a phenoxide anion. This anion then initiates an S_N2 reaction at the electrophilic carbon of the chloroether.

Recognise that this is not the final product, but is closely related to it.

Not needed here.

Write down the structure of the final product.

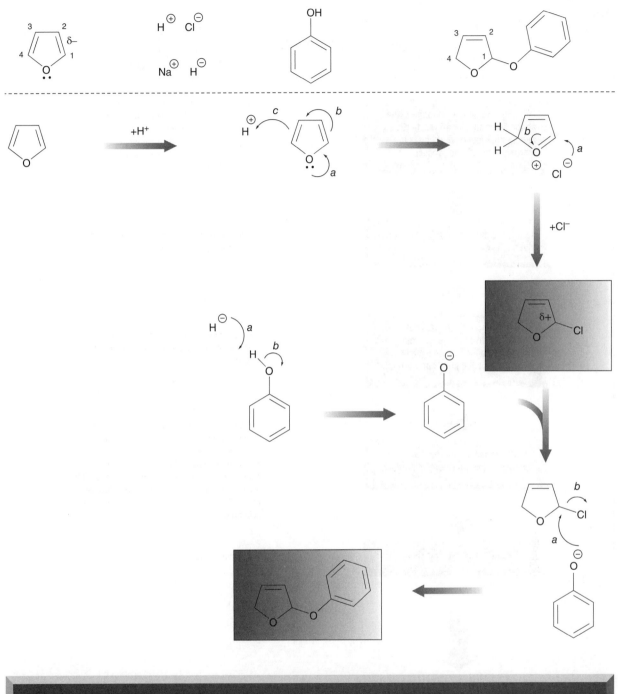

Summary: This is an example of electrophilic addition to activated alkenes:

Now try question 2.20

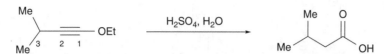

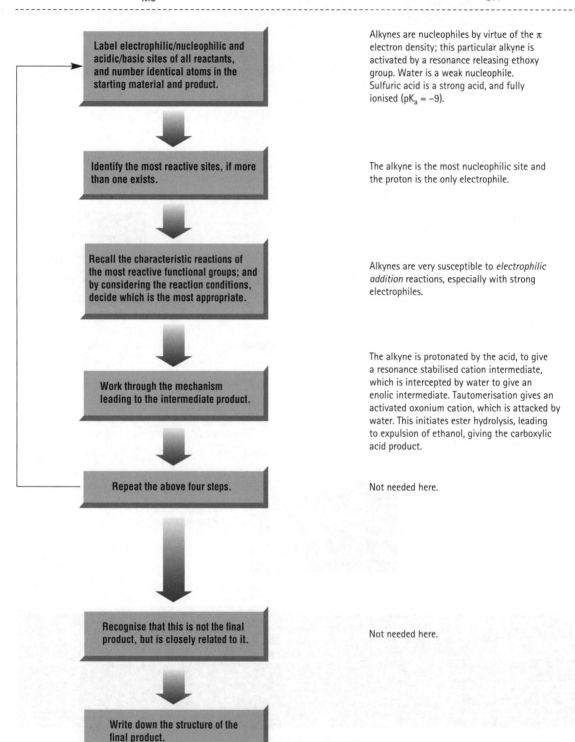

Label electrophilic/nucleophilic and acidic/basic sites of all reactants, and number identical atoms in the starting material and product.	Alkynes are nucleophiles by virtue of the π electron density; this particular alkyne is activated by a resonance releasing ethoxy group. Water is a weak nucleophile. Sulfuric acid is a strong acid, and fully ionised ($pK_a = -9$).
Identify the most reactive sites, if more than one exists.	The alkyne is the most nucleophilic site and the proton is the only electrophile.
Recall the characteristic reactions of the most reactive functional groups; and by considering the reaction conditions, decide which is the most appropriate.	Alkynes are very susceptible to *electrophilic addition* reactions, especially with strong electrophiles.
Work through the mechanism leading to the intermediate product.	The alkyne is protonated by the acid, to give a resonance stabilised cation intermediate, which is intercepted by water to give an enolic intermediate. Tautomerisation gives an activated oxonium cation, which is attacked by water. This initiates ester hydrolysis, leading to expulsion of ethanol, giving the carboxylic acid product.
Repeat the above four steps.	Not needed here.
Recognise that this is not the final product, but is closely related to it.	Not needed here.
Write down the structure of the final product.	

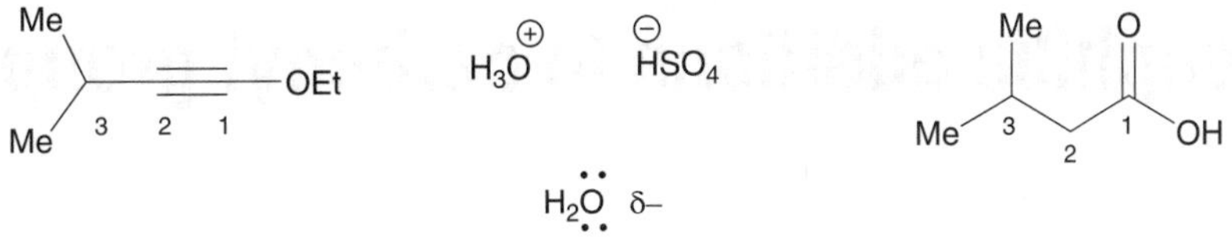

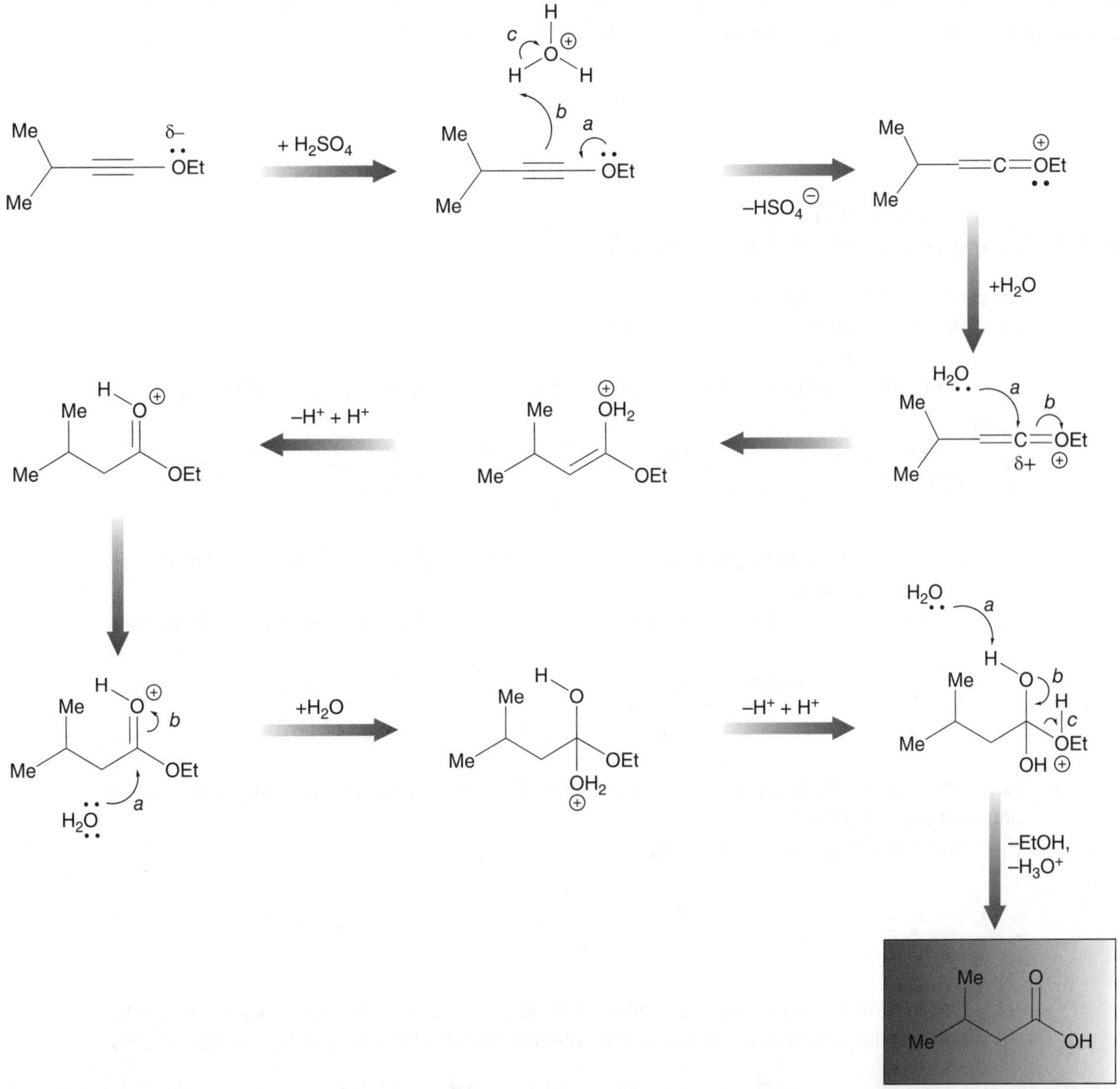

Me — C≡C — OEt   (3 2 1) $H_3O^{\oplus}$ $HSO_4^{\ominus}$ Me — CH — CH$_2$ — C(=O) — OH (3 2 1)

$H_2\overset{..}{\underset{..}{O}}$ δ−

+ H₂SO₄

−HSO₄⁻

+H₂O

−H⁺ + H⁺

+H₂O

−H⁺ + H⁺

−EtOH,
−H₃O⁺

3 Nucleophilic additions to carbonyl groups

Nucleophilic addition is a typical reaction of all carbonyl groups because of the polarisation of the C–O bond caused by the difference in electronegativity of the carbon and oxygen atoms:

Typical reactions depend on the nature of X, and are therefore:

Aldehydes and ketones (X = H, R respectively)

(a) Irreversible addition of nucleophiles.
   (i) Reduction. Addition of hydrogen, with
   - H_2 and a catalyst
   - 'H$^-$' from various hydride ion sources, such as complex metal hydrides (LiAlH$_4$ and NaBH$_4$)

   - 'Organic' hydride sources, such as the Cannizzaro reaction, and the Meerwein Ponndorf–Verley reaction
   (ii) Carbanion addition. Addition of Grignard reagents, organolithium reagents and acetylide anions.

(b) Reversible addition of nucleophiles. There are three types of carbonyl additions, depending on the nature of the nucleophile:
   (i) Addition. Water, bisulphite, cyanide.

$X = O, SO_3, CN$

   (ii) Addition–substitution. This mechanism is important for Group 6 nucleophiles, (i.e. oxygen and sulphur) leading to their respective products (hemiacetals and acetals, and thioacetals).

$X = O, S$

   (iii) Addition–elimination. This mechanism is important for Group 5 nucleophiles, (i.e. ammonia, hydroxylamine, hydrazine, semicarbazide, 2,4-dinitrophenylhydrazine) to form their respective products (imines, oximes, hydrazones, semicarbazones, 2,4-dinitrophenylhydrazones respectively).

How to Solve Organic Reaction Mechanisms: A Stepwise Approach, First Edition. Mark G. Moloney.
© 2015 John Wiley & Sons, Ltd. Published 2015 by John Wiley & Sons, Ltd. Companion website: www.wiley.com/go/moloney/mechanisms

If the amine is secondary, then enamine formation becomes possible, and enamines are important nucleophiles.

The Wittig reaction is a special case using a phosphorus ylide as the nucleophile.

Acyl compounds (X=OH (acids), X=OR (esters), X=NR$_2$ (amides), X=Cl, Br (acid halides) and acid anhydrides (X=OC(O)R)

Unlike aldehydes and ketones, in these acyl derivatives, X is a leaving group (a good leaving group is the conjugate base of a strong acid), and so the typical reactivity of any carbonyl group, that of nucleophilic addition, is almost invariably followed by elimination. The overall process is therefore an addition–elimination reaction, *via* a tetrahedral intermediate, which regenerates the carbonyl group after expulsion of the leaving group X$^-$.

The reactivity order for nucleophilic additions to carbonyl groups is:

$$RCOCl, RC(O)O(O)R > RCHO > R_2CO > RCO_2R' > RCONR'_2$$

This order of reactivity is dictated by the steric and electronic effects of the substituent X.

Typical reactions include:

(c)   Reduction. Addition of 'H$^-$' from hydride ion sources, such as complex metal hydrides (LiAlH$_4$).

(d)   Addition of hydroxide.

(e)   Nucleophiles. Addition of Grignard reagents, organolithium reagents and acetylide anions.

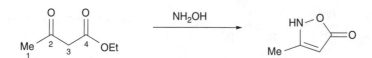

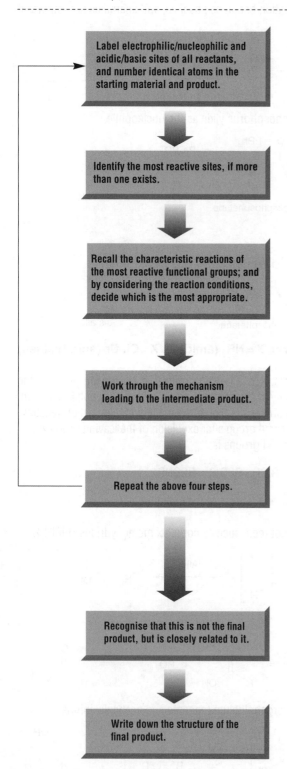

The carbonyl group is polarised by the *electronegativity* difference between carbon and oxygen, making the carbon electrophilic. Hydroxylamine is nucleophilic, since both nitrogen and oxygen atoms possess lone pairs.

Ketones are more reactive than esters due to the stabilising resonance delocalisation in the ester, and the nitrogen of hydroxylamine is more nucleophilic than oxygen, since nitrogen is less electronegative.

Hydroxylamine and ketones react to form *oximes* with *elimination* of water.

Oxime formation proceeds at a ketone by a standard *addition-elimination* mechanism with hydroxylamine and elimination of water.

The electrophilic ester carbonyl is susceptible to *nucleophilic addition*; the most nucleophilic site now available is the oxygen of the oxime. Alcohols and esters react to give esters, and here there is lactone (i.e. a cyclic ester) formation by standard *addition-elimination* mechanism at an ester.

The intermediate product can be converted to the final product by *tautomerisation*. This can be catalysed by any available acid or base.

Summary: Carbonyl compounds readily undergo addition–elimination reactions:

Now try questions 3.8 and 3.9

3.2

(i) CH₃Li (2 equiv.)
(ii) HCl, H₂O

Label electrophilic/nucleophilic and acidic/basic sites of all reactants, and number identical atoms in the starting material and product.

The hydroxylic hydrogen of a carboxylic acid is especially acidic, giving a resonance stabilised carboxylate anion. Methyl lithium is a carbanion equivalent, and is a powerful base and nucleophile.

Identify the most reactive sites, if more than one exists.

The carboxyl goup is the only reactive electrophilic site here. Methyl lithium is the only reactive nucleophile.

Recall the characteristic reactions of the most reactive functional groups; and by considering the reaction conditions, decide which is the most appropriate.

Acid-base proton exchange will always occur in preference to nucleophilic attack if both are possible; here the carboxylic acid is initially deprotonated by MeLi, generating the carboxylate anion and methane gas. However, MeLi is a sufficiently powerful nucleophile that is able to undergo *nucleophilic addition* to the carboxylate anion, even though it is already negatively charged.

Work through the mechanism leading to the intermediate product.

Nucleophilic addition of methyl anion to the carbonyl group generates a stable tetrahedral intermediate, which gives a ketal after acid treatment.

Repeat the above four steps.

The product is a ketal, which will undergo dehydration, by *elimination* of water under aqueous acidic conditions; the loss of water is assisted by one of the lone pairs of the ketal oxygens. Deprotonation of the *oxonium* ion gives the product.

Recognise that this is not the final product, but is closely related to it.

Not needed here.

Write down the structure of the final product.

Summary: Carboxylic acids react with 2 equivalents of an organolithium reagent to give a ketone:

$$RCO_2H \ + \ 2 \ R'Li \xrightarrow{\text{Acidic work-up}} \underset{R}{\overset{O}{\underset{\parallel}{C}}} R'$$

Now try questions 3.10 and 3.16

3.3

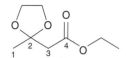

(i) 2 PhMgBr, then acidic work-up
(ii) HCl, H₂O

Label electrophilic/nucleophilic and acidic/basic sites of all reactants, and number identical atoms in the starting material and product.

The carbonyl group is polarised by the *electronegativity* difference between carbon and oxygen, making the carbon electrophilic. Phenylmagnesium bromide is a *carbanion* equivalent.

Identify the most reactive sites, if more than one exists.

The ester function is the only reactive electrophilic site, as the acetal is inert to all conditions except acidic *hydrolysis*. Phenylmagnesium bromide is the only nucleophile.

Recall the characteristic reactions of the most reactive functional groups; and by considering the reaction conditions, decide which is the most appropriate.

Esters react with excess Grignard reagent to give a tertiary alcohol in which two groups are identical.

Work through the mechanism leading to the intermediate product.

Nucleophilic *addition–elimination* to the ester carbonyl of phenylmagnesium bromide occurs, with alkoxide as the leaving group, giving a phenyl ketone. This is still electrophilic (carbonyl polarised by electronegativity difference between C and O) and a second addition of PhMgBr then occurs, which after protonation on work-up gives a *tertiary alcohol*.

Repeat the above four steps.

* Under aqueous acidic conditions, the acetal is readily *hydrolysed*, by the usual protonation-elimination-addition of water-elimination sequence for acetal hydrolysis.
* Tertiary alcohols are easily dehydrated under acidic conditions by *elimination* of water.

Recognise that this is not the final product, but is closely related to it.

Not needed here.

Write down the structure of the final product.

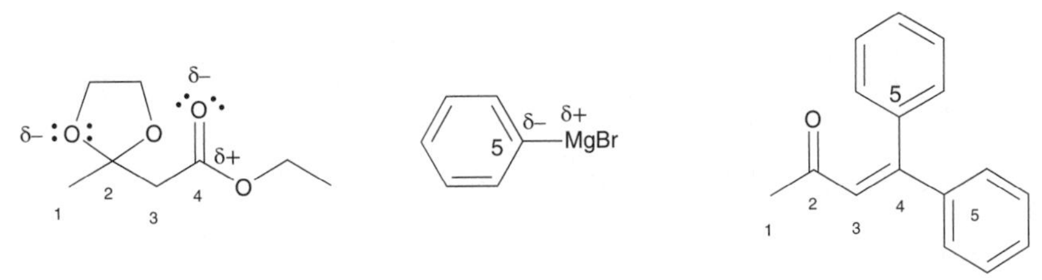

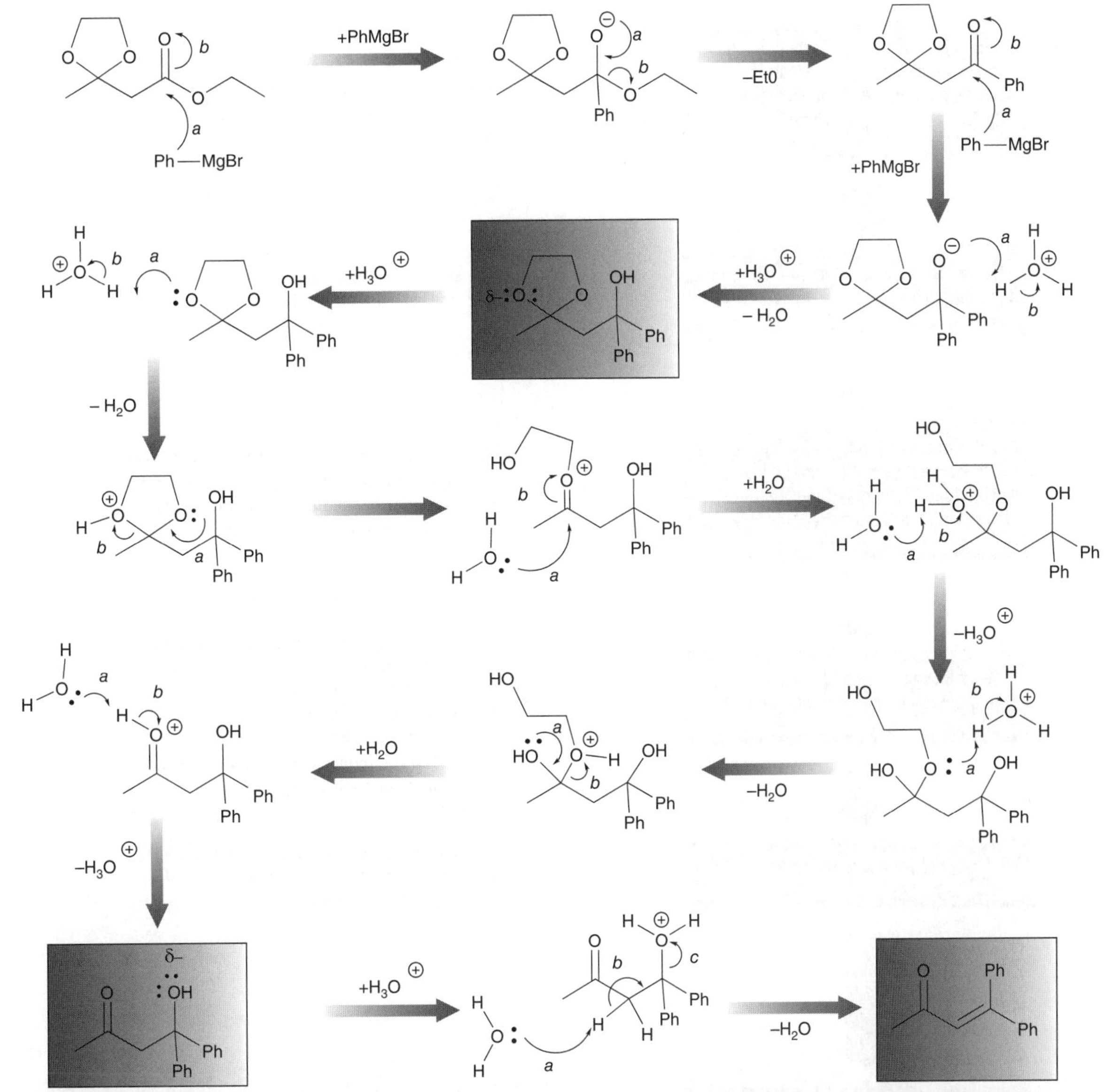

δ– :Ö: δ– :Ö:
2 4 δ+
1 3 δ+

δ– 5 δ+ MgBr

5
2 4 5
1 3

+PhMgBr

−EtO

+PhMgBr

+H₃O⁺
− H₂O

+H₃O⁺

+H₂O

− H₂O

−H₂O

−H₃O⁺

−H₂O

+H₂O

− H₂O

−H₃O⁺

+H₃O⁺

−H₂O

Ph—MgBr

Summary: An ester reacts with 2 equivalents of Grignard to give a tertiary alcohol:

$$RCO_2R' \; + \; 2 \, R''MgBr \longrightarrow \begin{array}{c} OH \\ | \\ R-C-R'' \\ | \\ R' \end{array}$$

Now try questions 3.11 and 3.17

3.4

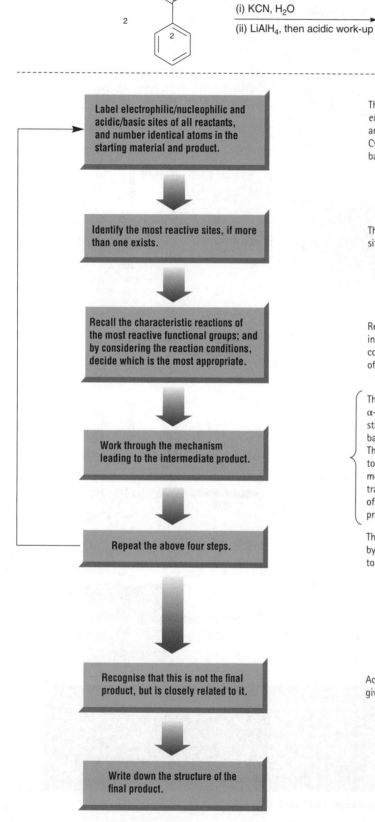

The carbonyl group is polarised by the *electronegativity* difference between carbon and oxygen, making the carbon electophilic. Cyanide is a very good nucleophile and weak base.

The aldehyde is the only electrophilic site, and cyanide is the only nucleophile.

Reversible addition of cyanide (NC⁻) to a carbonyl iniitally gives a cyanohydrin product; under these conditions, further addition to another molecule of the aldehyde starting material occurs.

The cyanohydrin product has a very acidic α-hydrogen (due to an inductively and resonance stabilised anion) and this is easily removed by base generated in the course of the reaction. This generates a nucleophile which is able to undergo *nucleophilic addition* to another molecule of benzaldehyde. After proton transfers, this intermediate collapses with loss of cyanide to give a hydroxyketone (benzoin) product.

The ketone product is readily reduced by LiAlH₄, by nucleophilic addition of hydride anion (H⁻), to give an alkoxide product.

Acidic work-up protonates the alkoxide product to give the 1,2-diol product (hydrobenzoin).

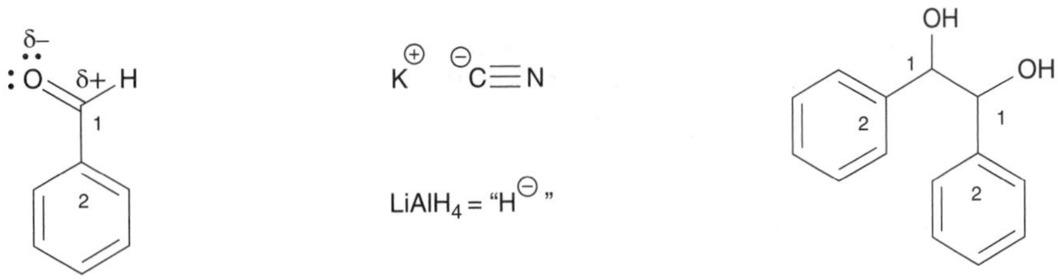

δ−

:O:

δ+ H

1

2

K⊕ ⊖C≡N

LiAlH₄ = "H⊖"

OH

OH

1

1

2

2

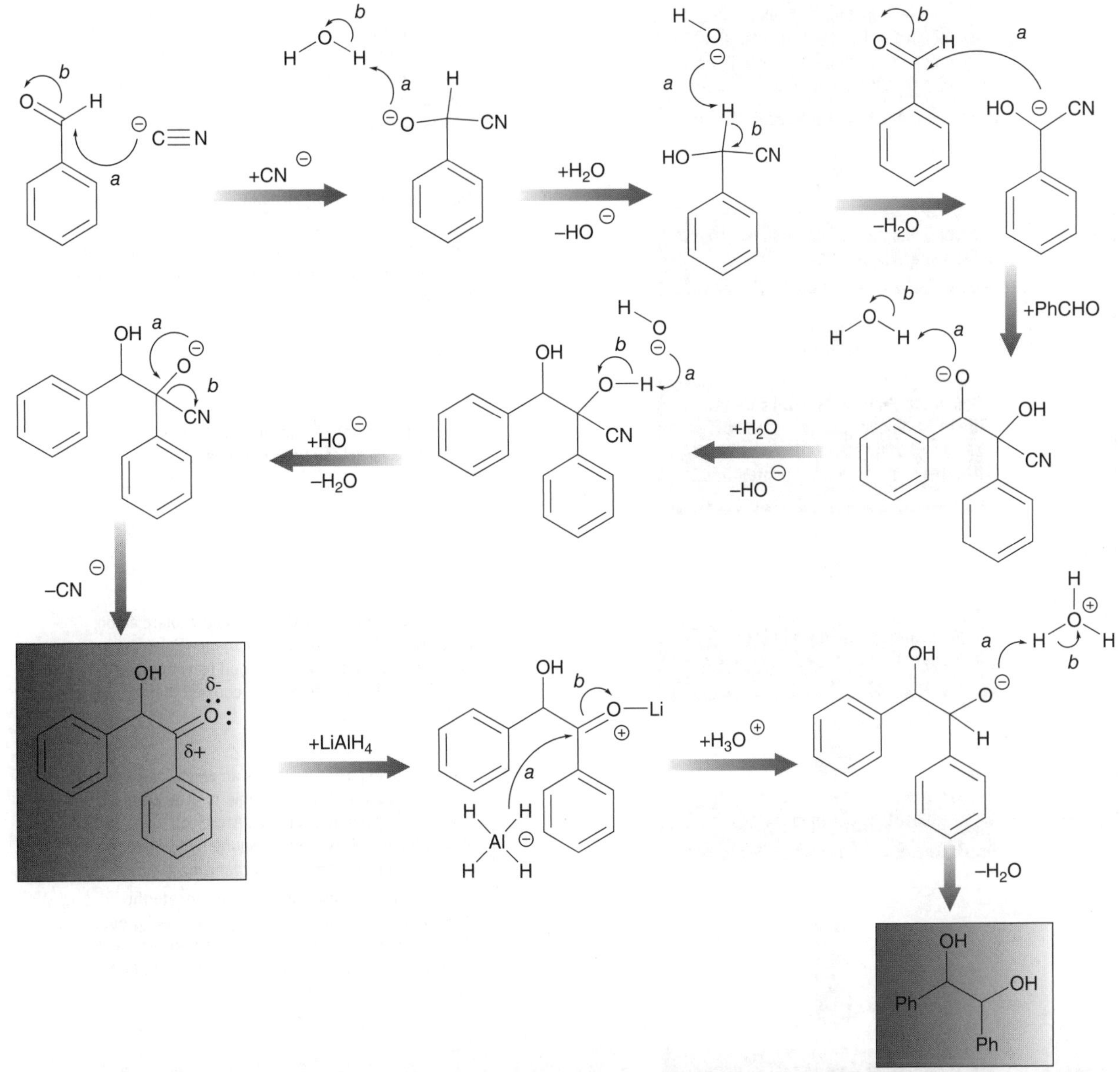

+CN⊖

b

a

+H₂O

−HO⊖

a

b

+PhCHO

b

a

−H₂O

+H₂O

−HO⊖

b

a

+HO⊖

−H₂O

a

b

−CN⊖

OH

δ−
:O:
δ+

+LiAlH₄

OH

O−Li⊕

a

b

H H

Al⊖

H H

+H₃O⊕

OH

O⊖

H

a

H O⊕ H
H
b

−H₂O

OH

OH

Ph

Ph

Summary: This is an example of the Benzoin condensation:

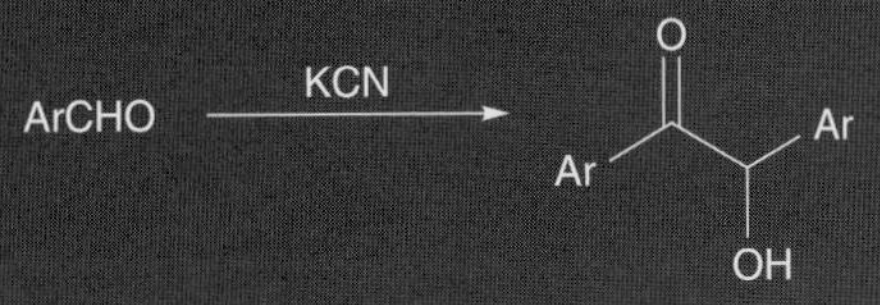

ArCHO —KCN→

O

Ar

Ar

OH

Now try questions 3.12 and 3.18

73

3.5

Label electrophilic/nucleophilic and acidic/basic sites of all reactants, and number identical atoms in the starting material and product.

α, β-Unsaturated ketones are electrophilic at both the carbon of the carbonyl group and at the β-position (here labelled C-1), by virtue of the *electronegativity* difference of the carbon and oxygen atoms.
HOOH is a good nucleophile (both oxygens possess lone pairs) and has a weak O–O bond. Sodium hydroxide is a good base.

Identify the most reactive sites, if more than one exists.

Nucleophiles preferentially add to the β-position of an α, β-unsaturated ketone, which is therefore a suitable electrophile. NaOH and HOOH set up an *equilibrium* which generates hydroperoxide anion, which is highly nucleophilic.

Recall the characteristic reactions of the most reactive functional groups; and by considering the reaction conditions, decide which is the most appropriate.

Conjugate addition of nucleophiles to an α, β-unsaturated ketone proceeds readily.

Work through the mechanism leading to the intermediate product.

Conjugate addition of HOO⁻ gives an enolate anion, which is itself also a good nucleophile. Since the hydroperoxide itself carries a good leaving group (OH), internal attack to give an epoxide then occurs.

Repeat the above four steps.

* The product epoxy ketone has an electrophilic carbonyl group, and is susceptible to attack by TsNHNH₂ (a good nucleophile). Attack of tosylhydrazine gives the corresponding hydrazone by an *addition–elimination* mechanism.

* The base HO⁻ is easily able to deprotonate the N–H, which is acidified by the adjacent tosyl group. Opening of the epoxide and fragmentation of the resulting intermediate gives the observed product directly.

Recognise that this is not the final product, but is closely related to it.

Not needed here.

Write down the structure of the final product.

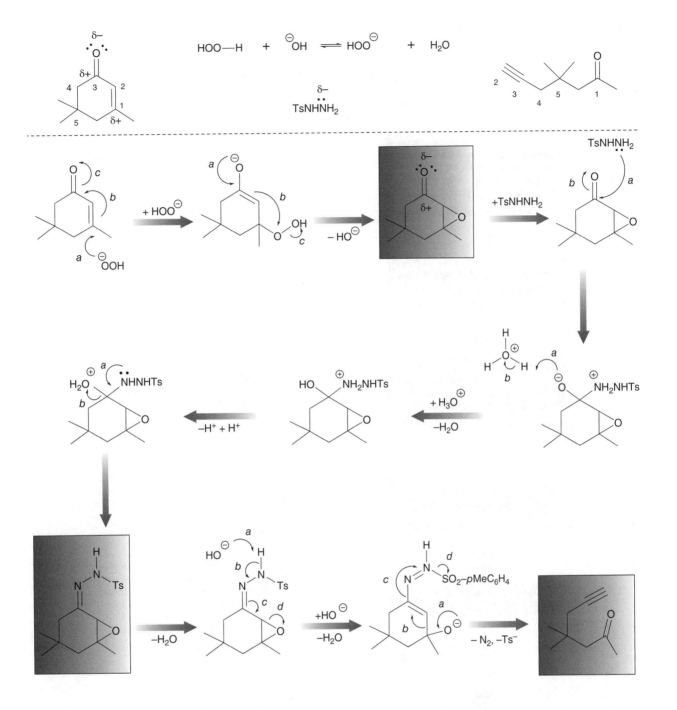

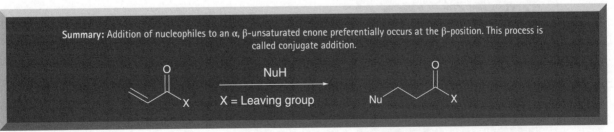

NuH

X = Leaving group

Now try questions 3.13 and 3.19

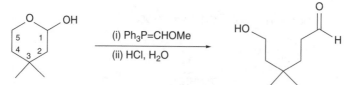

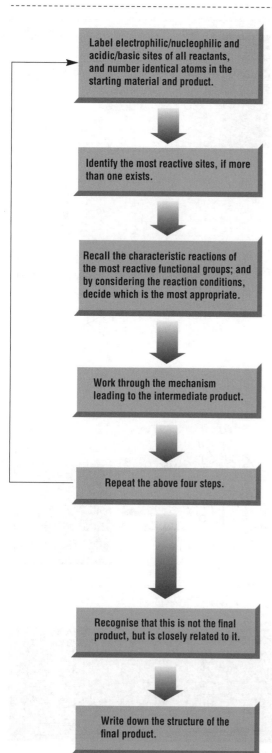

The starting material is a cyclic *hemiacetal*, which is in *equilibrium* with the corresponding hydroxy aldehyde, which is as usual electrophilic at the carbonyl carbon. Ylides are nucleophilic at the carbon adjacent to the P atom; this can be readily seen from the alternative resonance structure.

The aldehyde is the only electrophilic site, and the ylide the only nucleophile.

Carbonyl group and phosphorus ylides react in the characteristic *Wittig* reaction to give an alkene product.

Addition of the nucleophilic ylide to the carbonyl group is followed by *oxaphosphetane* formation. *Elimination* of $Ph_3P = O$ (very stable $P = O$ bond) then gives the alkene product, which in this case is an *enol ether*.

Enol ethers are both very basic and nucleophilic at the β-position and can be readily protonated there. Addition of water to the resulting intermediate cation gives another *hemiacetal*, which upon *elimination* of alcohol gives an oxonium ion. The product aldehyde results by deprotonation.

Not needed here.

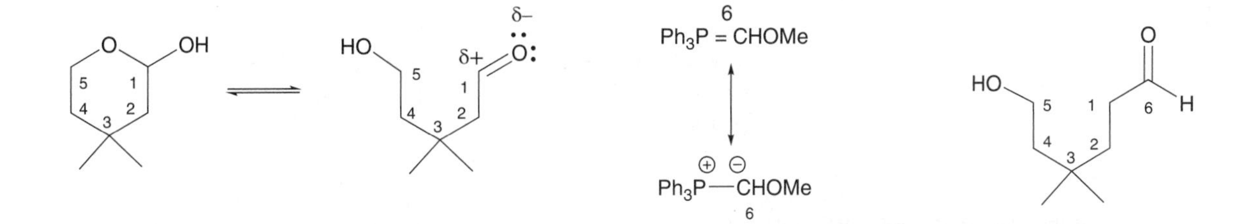

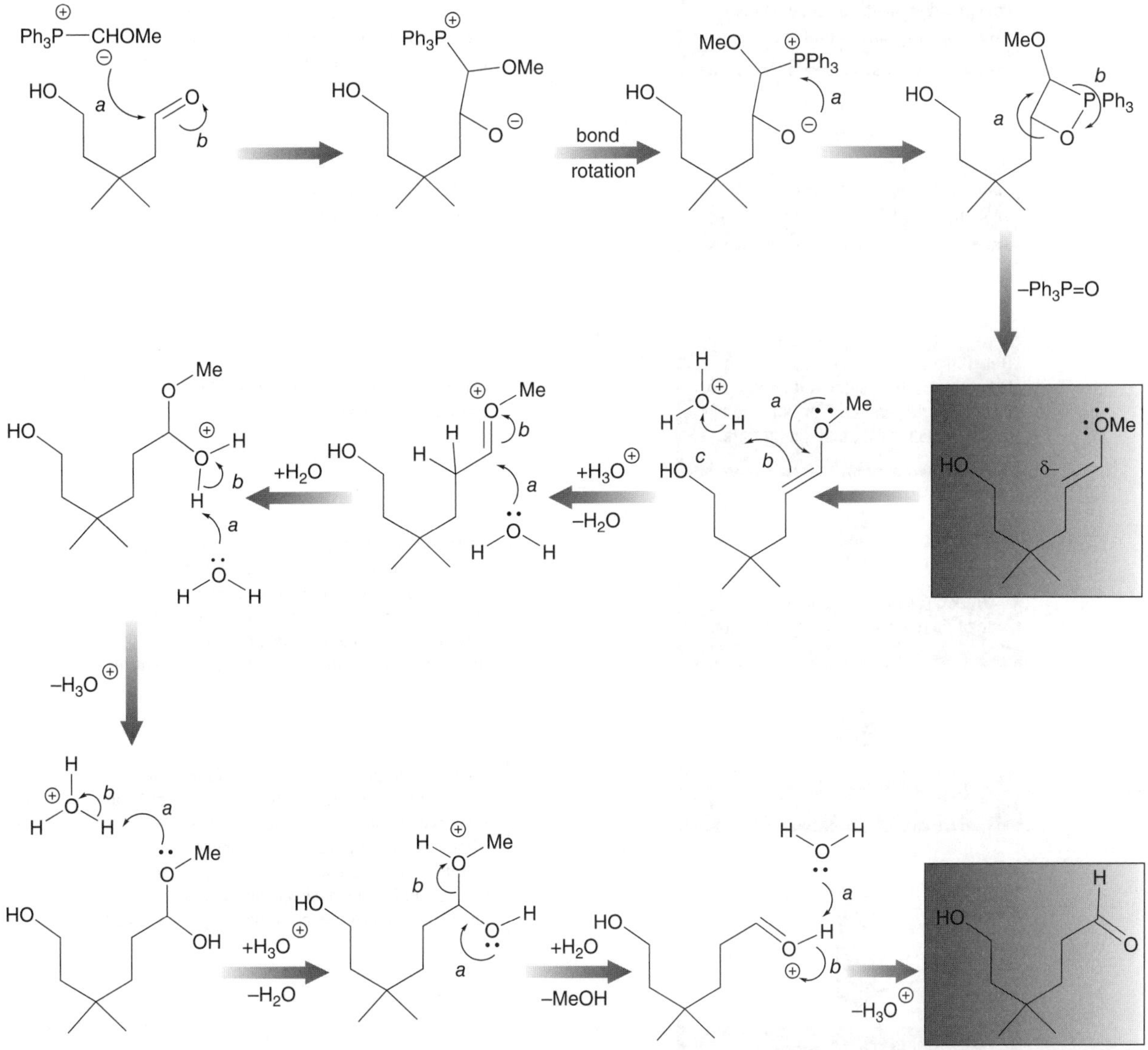

Ph₃P = CHOMe (labeled 6)

Ph₃P⁺—CHOMe⁻ (labeled 6)

bond rotation

−Ph₃P=O

+H₃O⁺
−H₂O

+H₃O⁺

+H₂O

−H₃O⁺

+H₃O⁺
−H₂O

+H₂O
−MeOH

−H₃O⁺

Summary: Carbonyl compounds readily react with phosphonium ylides to give alkenes (Wittig reaction):

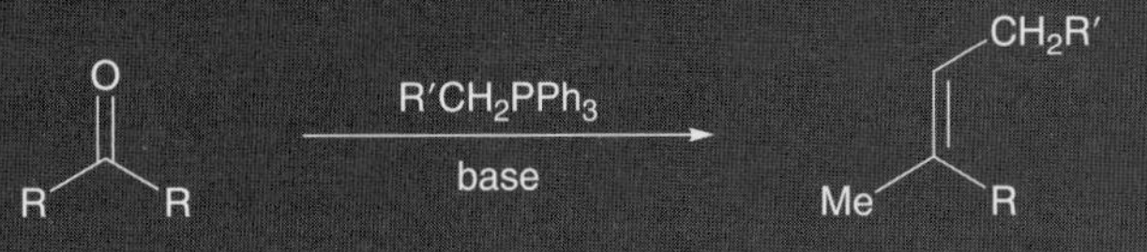

$$R_2C{=}O \xrightarrow[\text{base}]{R'CH_2PPh_3} \; \text{alkene}$$

Now try questions 3.14 and 3.20

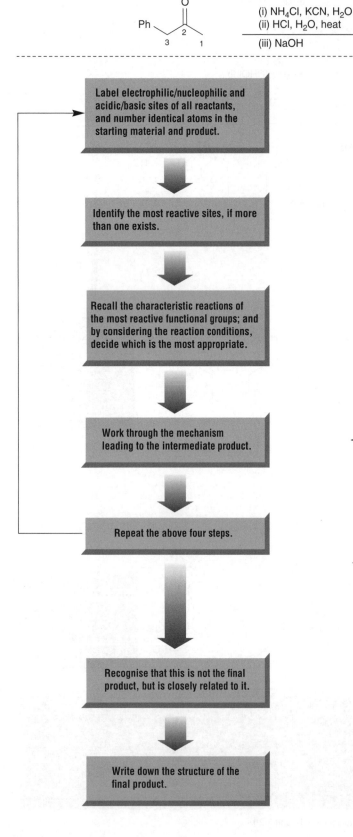

The carbonyl group is polarised by the *electronegativity* difference between carbon and oxygen, making the carbon electrophilic. Cyanide is a good nucleophile and a weak base. (sp hybridised carbon atom). Ammonium chloride generates ammonia in aqueous solution.

The ketone is the only electrophilic site, and ammonia and cyanide are both nucleophiles.

Ketones and amines react with elimination of water to give *imines*.

Imine formation proceeds by nucleophilic *addition-elimination* of NH_3 to the ketone carbonyl group, with loss of water. Imines, like ketones, are polarised by the *electronegativity* difference between carbon and nitrogen. *Nucleophilic addition* of cyanide occurs to give an amino nitrile intermediate.

* Acid catalysed hydrolysis of the nitrile to give an amide then occurs; this reaction proceeds by a sequence of addition–elimination steps.
* NaOH is a good base and nucleophile; *hydrolysis* of the amide function occurs under basic conditions by nucleophilic *addition–elimination* to give the corresponding acid product.

Not needed here.

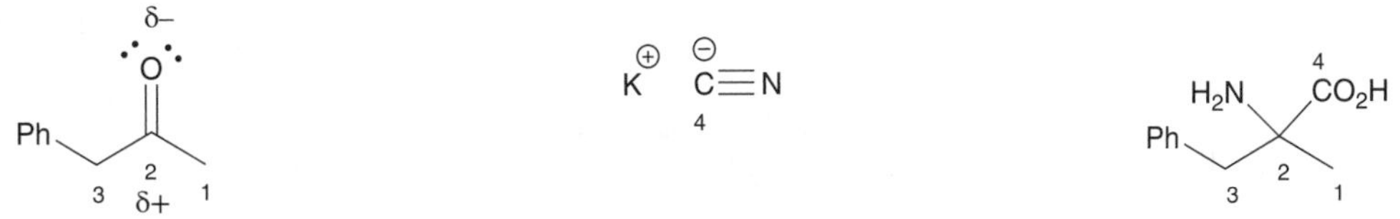

$$NH_4^+ + H_2O \rightleftharpoons NH_3 + H_3O^+$$

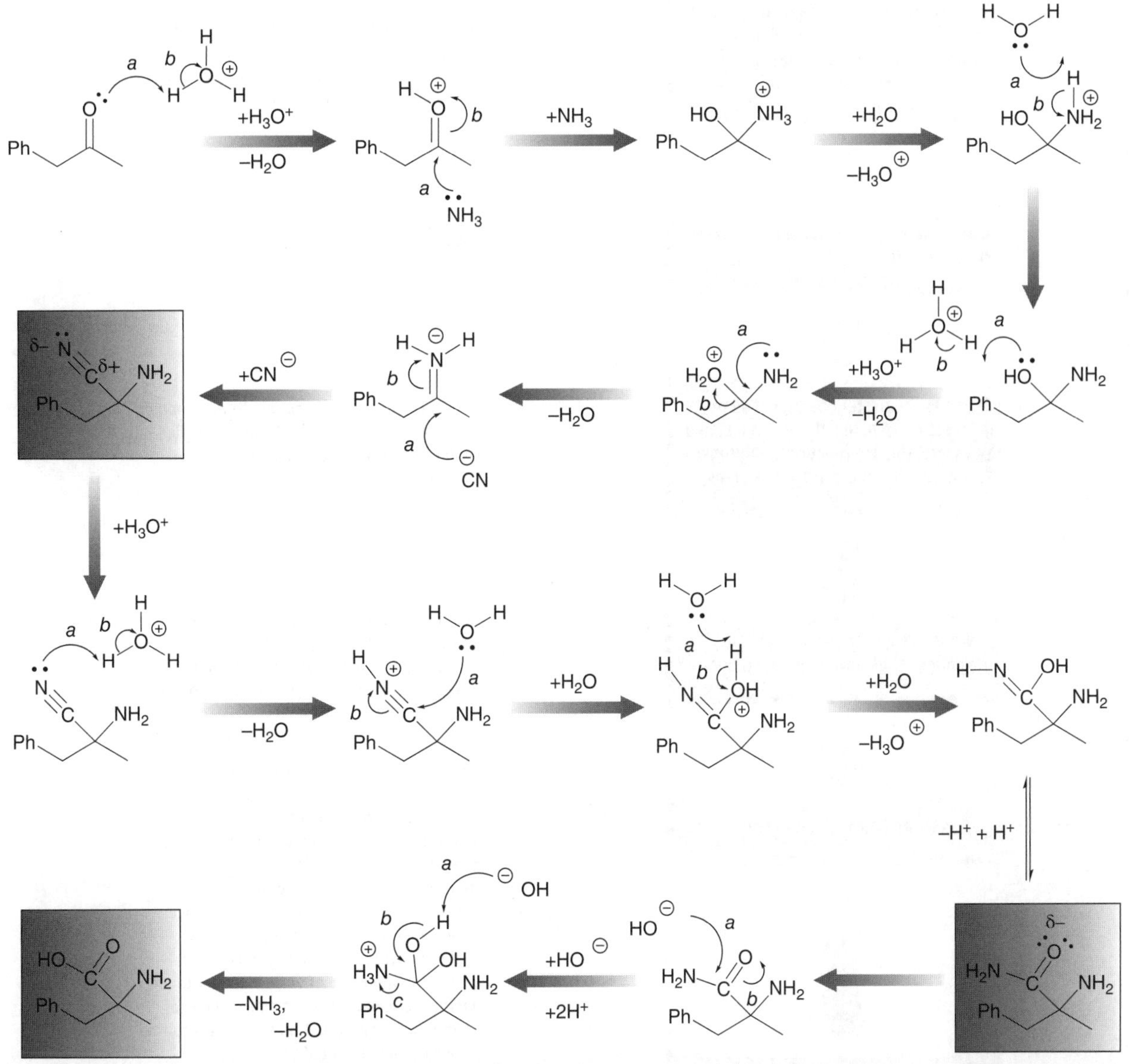

- -

+H$_3$O$^+$
−H$_2$O

+NH$_3$

+H$_2$O
−H$_3$O$^+$

+H$_3$O$^+$
−H$_2$O

−H$_2$O

+CN$^-$

+H$_3$O$^+$

−H$_2$O

+H$_2$O
−H$_3$O$^+$

−H$^+$ + H$^+$

+HO$^-$
+2H$^+$

−NH$_3$,
−H$_2$O

Summary: This is an example of the Strecker synthesis of amino acids:

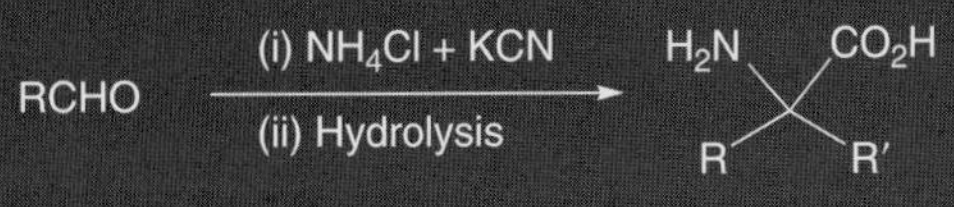

RCHO $\xrightarrow[\text{(ii) Hydrolysis}]{\text{(i) NH}_4\text{Cl + KCN}}$ H$_2$N—C(—CO$_2$H)(R)(R′)

Now try questions 3.15 and 3.21

3.8

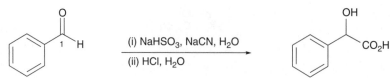

(i) NaHSO$_3$, NaCN, H$_2$O

(ii) HCl, H$_2$O

| Label electrophilic/nucleophilic and acidic/basic sites of all reactants, and number identical atoms in the starting material and product. | Aldehydes are electrophiles, since oxygen is electronegative and the double bond is reactive to nucleophiles. Sodium bisulfite is a weak acid; cyanide is a good nucleophile (sp hybridised carbon atom). |

| Identify the most reactive sites, if more than one exists. | The aldehyde is the only electrophilic site in the starting material. Cyanide is the only nucleophile. |

| Recall the characteristic reactions of the most reactive functional groups; and by considering the reaction conditions, decide which is the most appropriate. | Aldehydes are very susceptible to *nucleophilic addition* reactions, especially under acidic conditions in which they can be activated by protonation. |

| Work through the mechanism leading to the intermediate product. | The aldehyde is protonated by the weak acid bisulfite, and this activates it to direct attack by the highly nucleophilic cyanide anion. This leads to formation of a *cyanohydrin* product. |

| Repeat the above four steps. | * Hydrochloric acid is a strong acid; under the aqueous conditions, this leads to nitrile *hydrolysis*; protonation of the cyanide activates it to attack by water, which after a sequence of protonation-deprotonation steps leads to an amide product.
 * The amide product is still reactive under these conditions, and initial protonation further activates it to nucleophilic attack by another molecule of water, leading to an *addition-elimination*. Another sequence of protonation-deprotonation steps leads to the carboxylic acid product. |

| Recognise that this is not the final product, but is closely related to it. | Not needed here. |

| Write down the structure of the final product. | |

Na$^{\oplus}$ HSO$_3$$^{\ominus}$ H$^{\oplus}$ Cl$^{\ominus}$

K$^{\oplus}$ $^{\ominus}$CN   H$_2$O$^{\bullet\bullet}$ $\delta-$

Summary: Carbonyl compounds readily undergo addition and addition–elimination reactions:

$$\text{Nu} \xrightarrow[\text{X} \neq \text{Leaving group}]{\text{NuH}} \qquad \xrightarrow[\text{X = Leaving group}]{\text{NuH}} \text{Me–C(O)–Nu}$$

Now try question 3.9

3.9

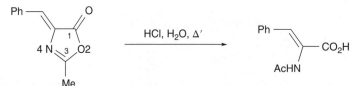

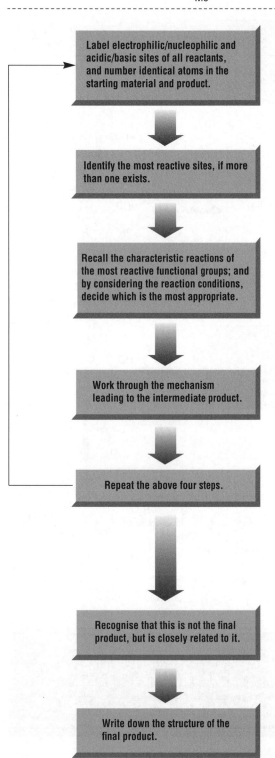

This oxazolone function has three possible electrophilic sites, but under acidic conditions is also capable of being protonated on the carbonyl group and amine groups. Hydrochloric acid is a strong acid, which under aqueous conditions generates a hydroxonium ion.

The imidate group is the most reactive site to protonation (N most basic), and this occurs on the nitrogen atom, giving an iminium cation intermediate.

Carbonyl species and their equivalents are very susceptible to *nucleophilic addition* and *nucleophilic addition-elimination* reactions, especially under aqueous conditions, leading to *hydrolysis*.

Such activated iminium species are very susceptible to *nucleophilic addition-elimination* reactions, which proceed by initial addition of the water nucleophile followed by a sequence of reversible proton transfers that lead to a tetrahedral intermediate which collapses with loss of carboxylate. This breaks open the ring and generates the product. Overall, the process is hydrolysis.

Not needed here.

Ph—CH= δ+ O (lone pairs)
1
:N 3 O 2 δ+
4 |
Me δ+

H₃O⊕ Cl⊖

H₂O δ−

Ph CH= 1 CO₂H
O
‖
Me—C 2 NH 3

Ph O
‖
4 :N 1
3 O 2
|
Me

+H₃O⊕
−H₂O

Ph O
:N⊕
H a
H—O—H b
⊕

Ph O
H—N⊕
b Me a
H—O—H

+H₂O

Ph O
H—N
Me OH₂ ⊕

+H₂O

Ph O
a :O—H
H—N b
Me O⊕ H
|
H

−H₃O⊕

Ph O
HN
Me OH

+H₃O⊕

Ph O
a :O:
HN H—O⊕—H
Me O—H b H

−H₂O

Ph c O⊕—H
HN b
Me O—H a

Ph CO₂H
O
‖
Me—C NH

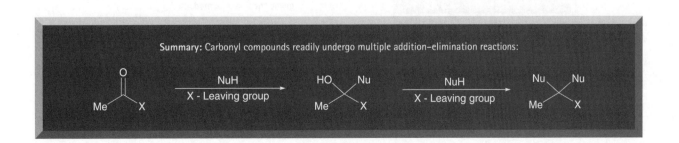

Summary: Carbonyl compounds readily undergo multiple addition−elimination reactions:

O
‖   NuH   HO   Nu   NuH   Nu   Nu
Me—C—X ────────→   \ /   ────────→   \ /
     X - Leaving group Me X   X - Leaving group Me X

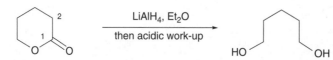

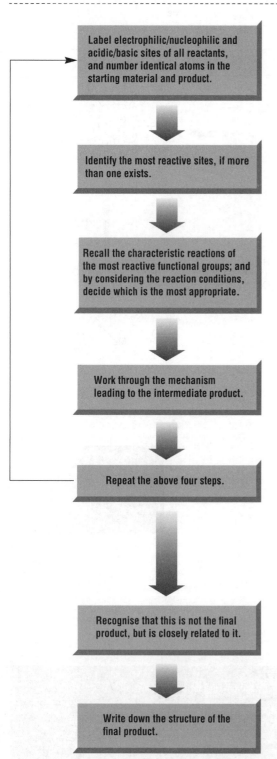

Label electrophilic/nucleophilic and acidic/basic sites of all reactants, and number identical atoms in the starting material and product.	*Lactones* (cyclic esters) are electrophilic at the carbonyl group, since oxygen is electronegative. Lithium aluminium hydride is a source of hydride, which is an excellent nucleophile, especially when the lithium cation acts as a Lewis acid to further activate the carbonyl group.
Identify the most reactive sites, if more than one exists.	The carbonyl group is the only electrophilic site, and lithium aluminium hydride the only nucleophile.
Recall the characteristic reactions of the most reactive functional groups; and by considering the reaction conditions, decide which is the most appropriate.	Carbonyl groups readily interact with Lewis acids to generate an activated oxonium ion, which activates the carbonyl group to nucleophilic addition by a suitable nucleophile.
Work through the mechanism leading to the intermediate product.	Addition of the hydride nucleophile onto the activated carbonyl group leads to the formation of a lithium alkoxide salt, which fragments further by loss of alkoxide and in so doing generates another aldehyde; this is in turn intercepted by another hydride, which gives a lithium bisalkoxide.
Repeat the above four steps.	Not needed here.
Recognise that this is not the final product, but is closely related to it.	Work-up proceeds by addition of weak aqueous acid. This protonates both alkoxides in two successive steps, giving the final diol product.
Write down the structure of the final product.	

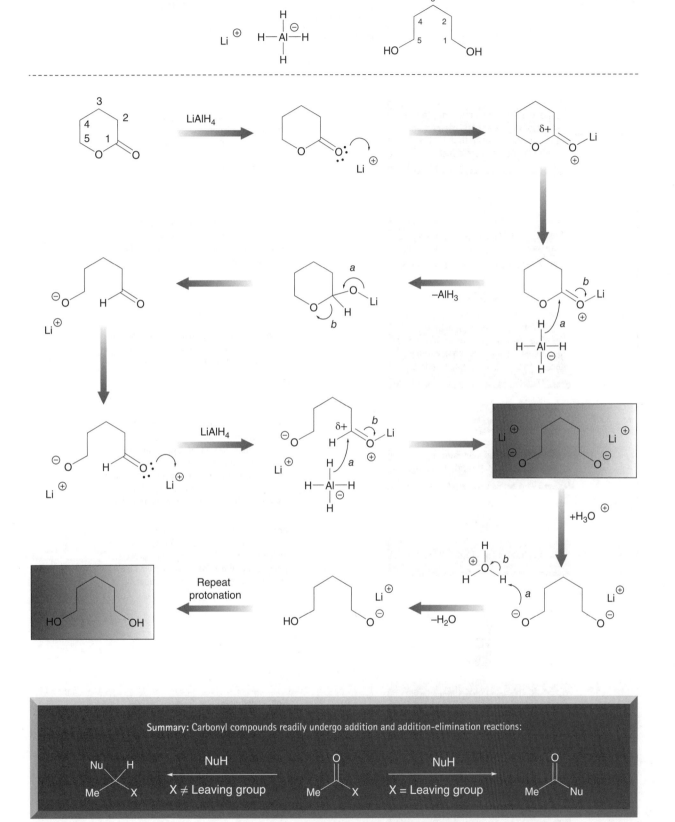

Summary: Carbonyl compounds readily undergo addition and addition-elimination reactions:

Now try question 3.16

85

3.11

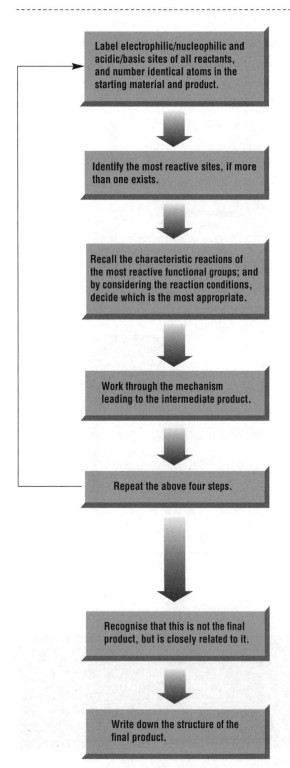

HO — (4) — (3) — (2) — CO₂Me
         |
        NH₂ (1)

$\xrightarrow{\text{(i) }^t\text{BuCHO, Et}_3\text{N}}$ (ii) CH₃COCl, Et₃N

[oxazolidine product with CO₂Me, t-Bu, N–C(=O)–Me (Ac) group]

Label electrophilic/nucleophilic and acidic/basic sites of all reactants, and number identical atoms in the starting material and product.

Amino esters have carbonyl groups that which are electrophiles, and amine and alcohol groups which are nucleophiles. Aldehydes are electrophiles, since oxygen is electronegative and the double bond is reactive to nucleophiles. Triethylamine is a weak base.

Identify the most reactive sites, if more than one exists.

The amine is the most nucleophilic site of the amino ester (since N is less electronegative than O) and the pivaldehyde carbonyl is the most reactive electrophile.

Recall the characteristic reactions of the most reactive functional groups; and by considering the reaction conditions, decide which is the most appropriate.

Amines and aldehydes react to form *imines*; this reaction can be conducted under acidic or basic catalysed conditions.

Work through the mechanism leading to the intermediate product.

The amino group adds to the electrophilic carbon of the aldehyde function, and proton transfer generates an aminal; elimination of water forms the corresponding imine. As a result of the proximity of the adjacent hydroxyl group, ring closure, followed by proton transfer, to form a cyclic oxazolidine will occur.

Repeat the above four steps.

Acetyl chloride is highly electrophilic, and converts the amine to the corresponding amide *via* an *addition–elimination* process. Deprotonation mediated by triethylamine leads to the formation of the product.

Recognise that this is not the final product, but is closely related to it.

Not needed here.

Write down the structure of the final product.

HO—CH₂ ... CO₂Me (starting materials row)

−H⁺ + H⁺

−H₂O   +Et₃N

−H⁺ + H⁺

+ Et₃N

−HNEt₃⁺Cl⁻

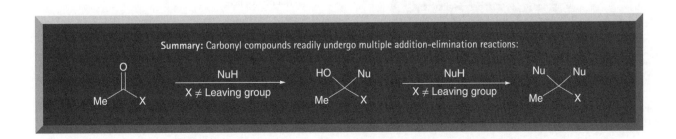

Summary: Carbonyl compounds readily undergo multiple addition–elimination reactions:

Me—C(=O)—X →[NuH / X ≠ Leaving group]→ HO—C(Me)(X)—Nu →[NuH / X ≠ Leaving group]→ Nu—C(Me)(X)—Nu

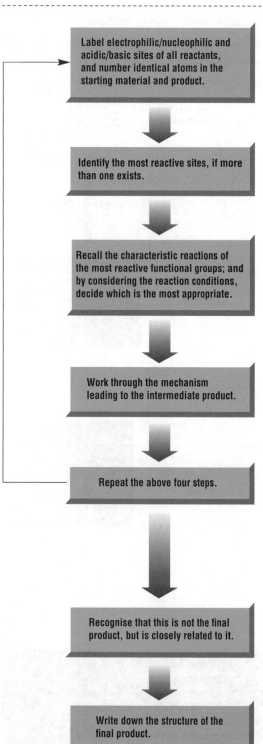

Aldehydes are electrophiles, since oxygen is electronegative and the double bond is reactive to nucleophiles. Potassium cyanide is an excellent nucleophile (sp hybridised carbon atom), but also a weak base.

The aldehyde is the only electrophilic site in the starting material, and cyanide is the only nucleophile.

Aldehydes are very susceptible to *nucleophilic addition* reactions, especially when the nucleophile is a good one like cyanide, but importantly these additions are reversible.

The aldehyde is directly attacked by the highly nucleophilic cyanide anion. This leads to formation of a cyanoalkoxide product, which under the conditions of the reaction, equilibrates by proton transfer to generate a resonance stabilised carbanion. This reacts with a second equivalent of the starting aldehyde to generate another alkoxide. Proton transfer then regenerates the original cyanoalkoxide intermediate, which collapses (cyanide addition to a carbonyl group is reversible) to regenerate the original carbonyl group.

Not needed here.

Now try question 3.18

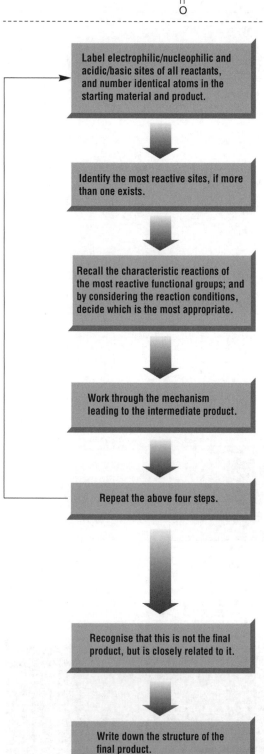

Aldehydes are electrophiles, since oxygen is electronegative and the double bond is reactive to nucleophiles. This substrate, glyoxal, is especially reactive as there are two adjacent aldehydes whose dipoles reinforce reactivity. Sodium hydroxide is an excellent base and nucleophile.

The aldehyde is the only electrophilic site in the starting material and of course both are equivalent. Hydroxide is the only base and nucleophile.

Aldehydes are very susceptible to *nucleophilic addition* reactions, especially when the nucleophile is a good one like hydroxide.

The aldehyde is directly attacked by the highly nucleophilic hydroxide anion. This leads to formation of an alkoxide product, which under the conditions of the reaction, undergoes a slow second addition of hydroxide to the other aldehyde carbonyl group to generate a dianion. Intramolecular hydride transfer with loss of hydroxide gives another alkoxide intermediate.

Not needed here.

Protonation on work up generates the product.

Summary: This is an example of the Cannizzaro reaction, although in this case hydride transfer is intramolecular:

$$2\ \text{ArCHO} \longrightarrow \text{ArCO}_2\text{H} + \text{ArCH}_2\text{OH}$$

Now try question 3.19

3.14

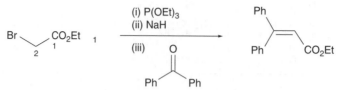

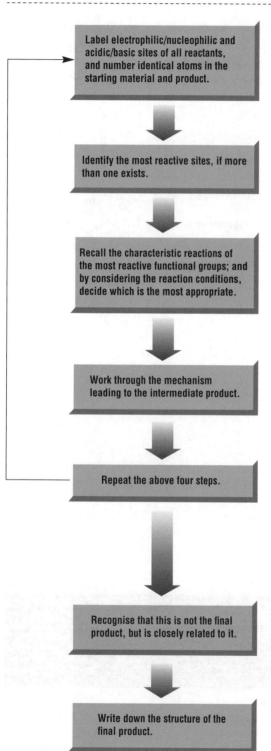

Label electrophilic/nucleophilic and acidic/basic sites of all reactants, and number identical atoms in the starting material and product.	α-Haloesters are excellent electrophiles, since the halogen is electronegative and a good leaving group. The ester carbonyl is also electrophilic. Triethyl phosphite is a very good nucleophile.
Identify the most reactive sites, if more than one exists.	The carbon adjacent to the bromine is the most electrophilic position, and phosphorus is the only nucleophile.
Recall the characteristic reactions of the most reactive functional groups; and by considering the reaction conditions, decide which is the most appropriate.	Alkyl halides are very susceptible to *nucleophilic substitution* reactions, and this one especially so due to the adjacent ester group which further activates nucleophilic attack.
Work through the mechanism leading to the intermediate product.	Triethyl phosphite displaces the bromine in an S_N2-like process, and back attack by the bromide which is released generates a phosphonate product, in which the α-protons are strongly acidic.
Repeat the above four steps.	Sodium hydride is a strong base; deprotonation of the α-protons proceeds easily, generating a phosphonate anion, which reacts with the added ketone by *nucleophilic addition*. Cyclisation to form an oxaphosphetane intermediate is followed by collapse with formation of a P=O bond giving the product alkene.
Recognise that this is not the final product, but is closely related to it.	Not needed here.
Write down the structure of the final product.	

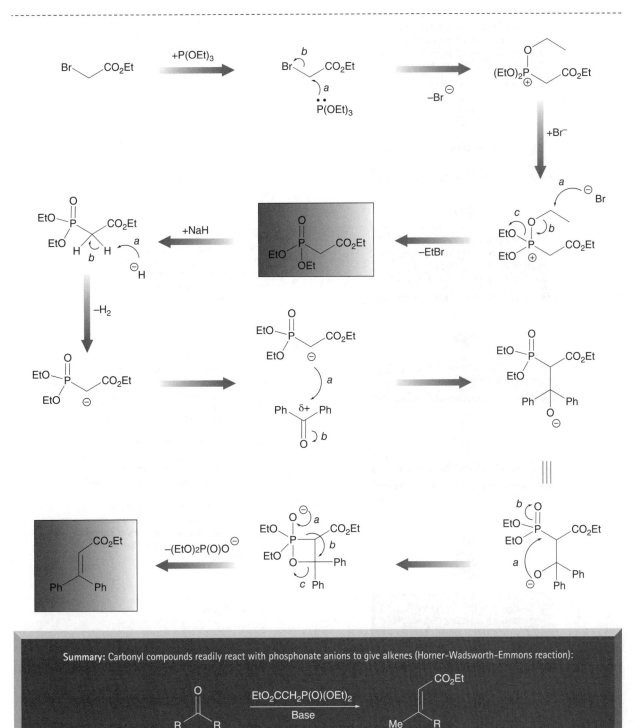

Summary: Carbonyl compounds readily react with phosphonate anions to give alkenes (Horner-Wadsworth-Emmons reaction):

Now try question 3.20

3.15

$$\underset{\substack{2 \quad 1}}{CH_3CHO} \quad \xrightarrow[\text{(ii) HCl, } H_2O]{\text{(i) NaCN, } NH_4Cl} \quad \underset{CO_2H}{\overset{H_3C \quad NH_2}{\diagdown \diagup}}$$

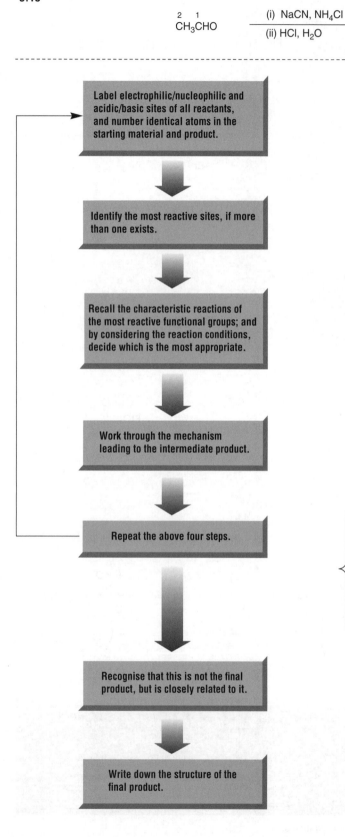

Label electrophilic/nucleophilic and acidic/basic sites of all reactants, and number identical atoms in the starting material and product.	Aldehydes are electrophiles, since oxygen is electronegative and the double bond is reactive to nucleophiles. Ammonium chloride is in equilibrium with ammonia, an excellent nucleophile; sodium cyanide is an excellent nucleophile (sp hybridised carbon atom).
Identify the most reactive sites, if more than one exists.	The aldehyde is the only electrophilic site in the starting material. Cyanide and ammonia are both nucleophiles. Cyanide reacts reversibly with aldehydes, so imine formation is preferred.
Recall the characteristic reactions of the most reactive functional groups; and by considering the reaction conditions, decide which is the most appropriate.	Aldehydes are very susceptible to *nucleophilic addition* reactions, especially in the presence of good nucleophiles like ammonia, to give imines. Under the conditions of this reaction, further addition by cyanide occurs.
Work through the mechanism leading to the intermediate product.	The aldehyde is attacked by the nucleophilic ammonia, which after a series of proton transfers and loss of water, leads to the formation of an imminium ion which is in turn intercepted by the nucleophile, cyanide. This leads to formation of a *aminonitrile* product.
Repeat the above four steps.	* Hydrochloric acid is a strong acid, which protonates the nitrile and activates it to attack by water. A series of proton transfers and finally tautomerisation then leads to the *amide* product. * Further reaction with hydrochloric acid protonates the carbonyl group and activates it to attack by water again. A series of proton transfers and finally elimination then leads to the *acid* product.
Recognise that this is not the final product, but is closely related to it.	Not needed here.
Write down the structure of the final product.	

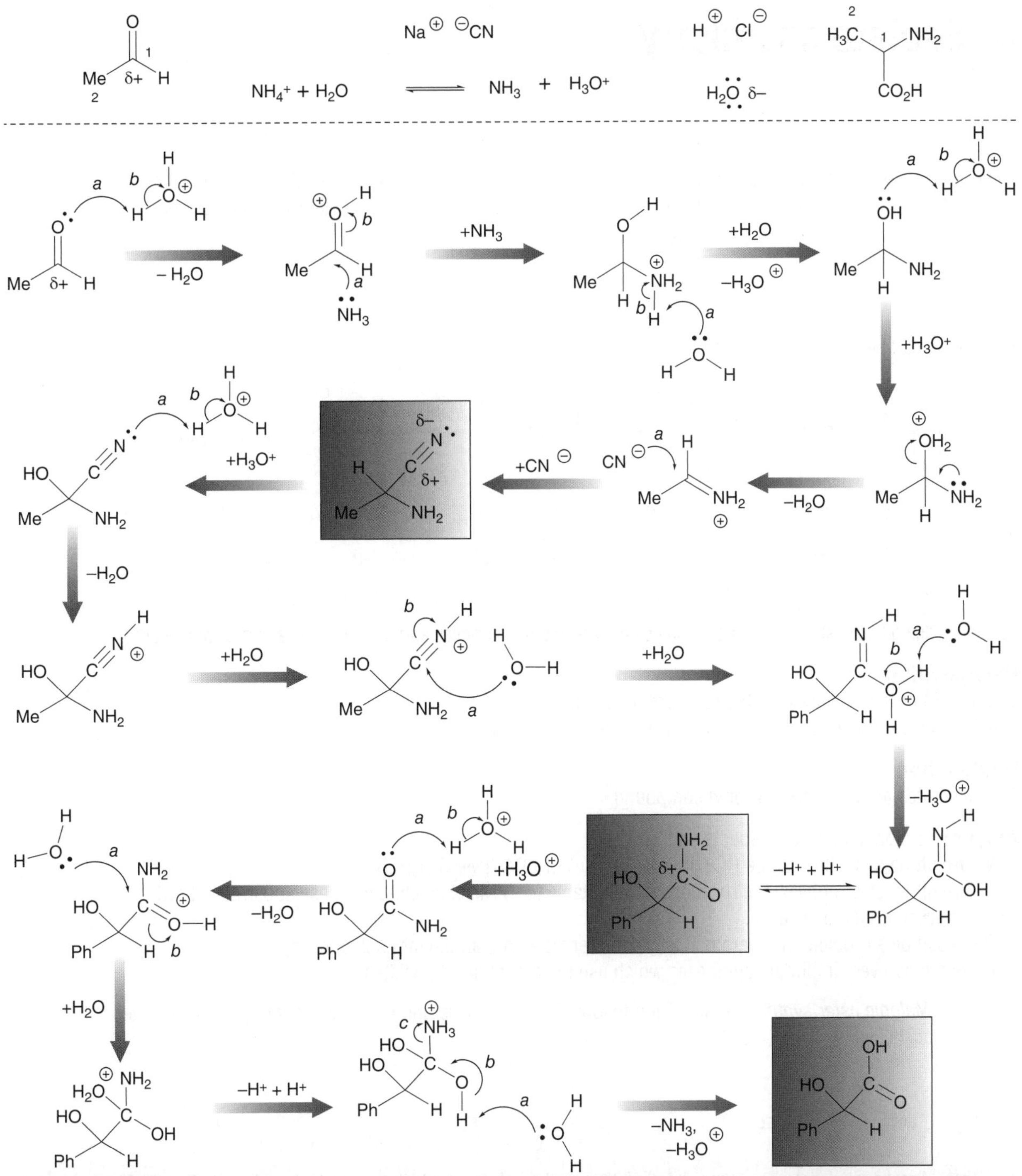

Summary: This is an example of the Strecker synthesis of amino acids:

$$RCHO \xrightarrow[\text{(ii) Hydrolysis}]{\text{(i) NH}_4\text{Cl + KCN}} \underset{R \quad R'}{H_2N \quad CO_2H}$$

Now try question 3.21

95

4 Enolate chemistry

The α-protons of carbonyl groups are acidic due to both inductive withdrawal of the adjacent carbonyl group, which weakens the carbon–hydrogen bond, and resonance stabilisation of the enolate which is formed.

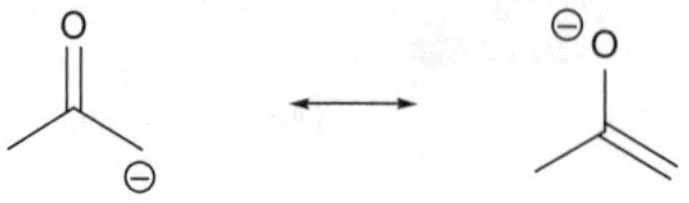

Enolisation can be acid or base catalysed:

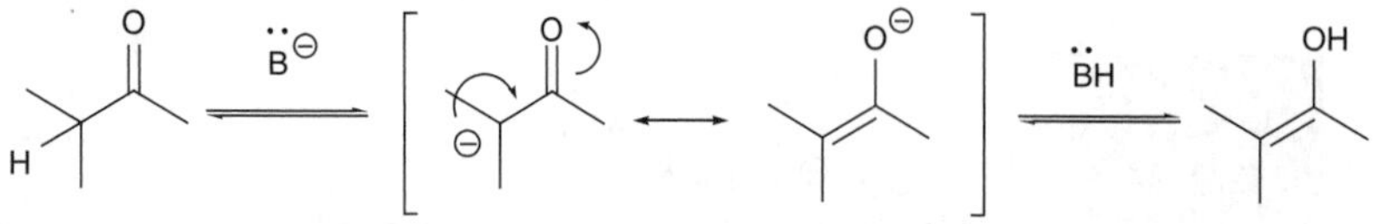

As well as being basic, such enols and enolates are excellent nucleophiles and participate in a variety of reactions:

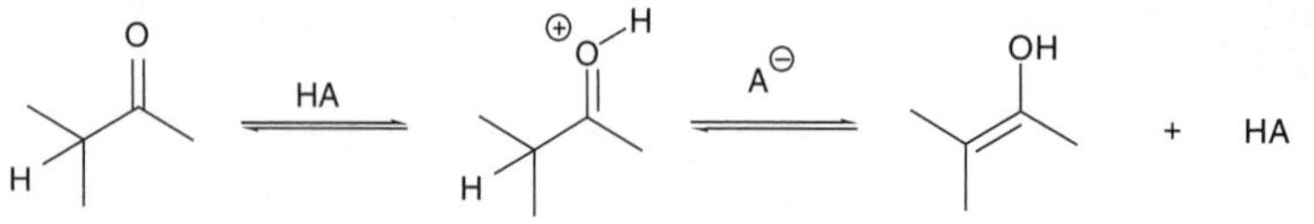

Halogenation
Can be acid or base catalysed—the Iodoform reaction.
Hell–Vollhardt–Zelinsky reaction is used to prepare α-haloacid halides.

Deuteriation
This generates an α-deuteriocarbonyl compound.

Alkylation. Some important aspects are:
Nature of the base: this can be RO⁻ (R = Me, Et, tBu) or LDA ([Me₂CH]₂NLi).
Nature of the electrophiles: the reaction is not suitable for 3° alkyl halides; for these substrates, Lewis acid–catalysed alkylation is used.
There can be a problem of O versus C alkylation (enolates are ambident nucleophiles).
There are two very important syntheses which use the alkylation of enolates:

(a) *Malonic ester synthesis*—alkylation followed by hydrolysis and decarboxylation gives substituted acetic acids:

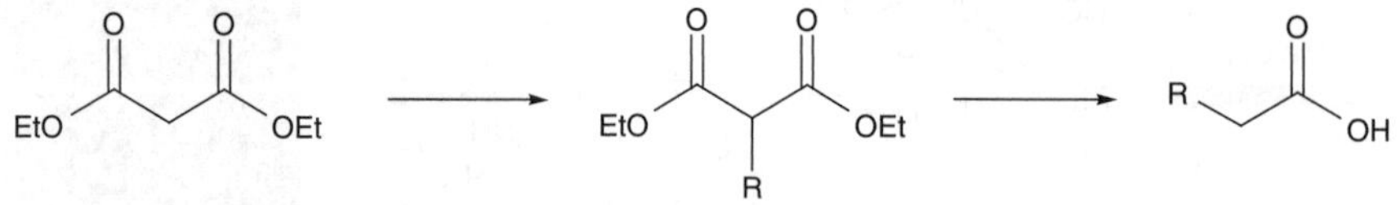

(b) *Acetoacetic ester synthesis*—alkylation followed by hydrolysis and decarboxylation gives substituted acetones:

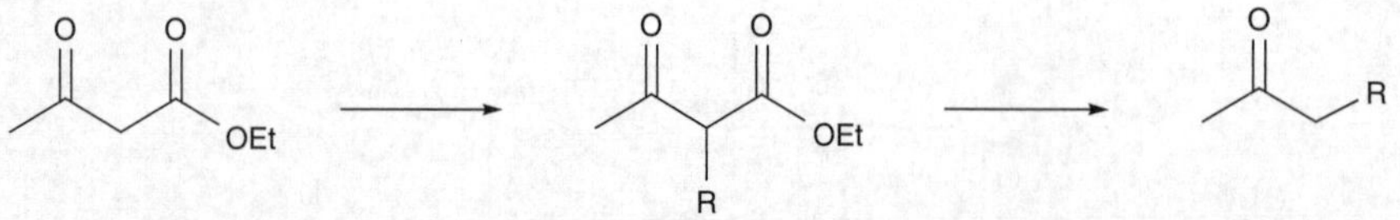

How to Solve Organic Reaction Mechanisms: A Stepwise Approach, First Edition. Mark G. Moloney.
© 2015 John Wiley & Sons, Ltd. Published 2015 by John Wiley & Sons, Ltd. Companion website: www.wiley.com/go/moloney/mechanisms

Condensation reactions. There are many examples of this process.

(a)   *Aldol condensation*
      Acid- and base-catalysed reactions.
      Mixed and intramolecular aldols.
      Dehydration to give α,β-unsaturated carbonyl derivatives.

(b)   *Knoevenagel condensation*

(c)   *Michael reaction and the Robinson Ring annulation*

(d)   *Claisen ester condensation* (and the cyclic variant, the Dieckmann cyclisation)

(e)   *Other condensations*: Perkin, Reformatsky, Stobbe, Darzens reactions.

97

4.1

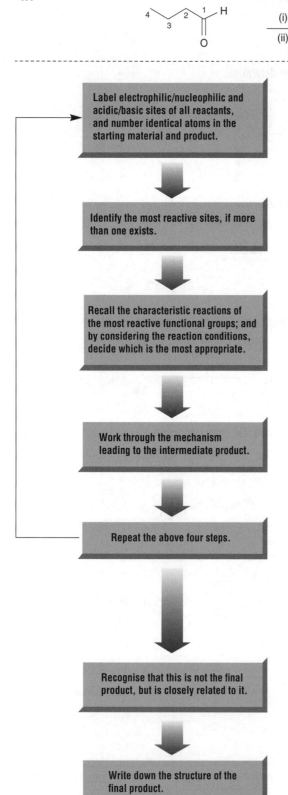

(i) KOH, H_2O

(ii) HCl, H_2O, Δ'

Label electrophilic/nucleophilic and acidic/basic sites of all reactants, and number identical atoms in the starting material and product.

The carbonyl group is polarised by the *electronegativity* difference between carbon and oxygen, making the carbon electrophilic. The α-protons of a carbonyl group are acidified by *inductive* withdrawl, and the resulting enolate is resonance stabilised and a good nucleophile. KOH is a good base.

Identify the most reactive sites, if more than one exists.

The aldehyde function is the only group capable of enolisation under basic conditions, and sodium hydroxide is the strongest base.

Recall the characteristic reactions of the most reactive functional groups; and by considering the reaction conditions, decide which is the most appropriate.

Enolisation of the aldehyde is followed by standard *nucleophilic addition* to another carbonyl group of the starting aldehyde; this is an example of the *aldol reaction.*

Work through the mechanism leading to the intermediate product.

Intermolecular *nucleophilic addition* gives condensation of two molecules of the aldehyde, to give a β–hydroxyaldehyde.

Repeat the above four steps.

The β-hydroxyaldehyde can be easily protonated on the hydroxyl oxygen by strong acid, and *elimination* to the α, β -unsaturated product occurs.

Recognise that this is not the final product, but is closely related to it.

Not needed here.

Write down the structure of the final product.

Summary: This is an example of an aldol reaction, that results in the condensation of two equivalents of an aldehyde to give a β-hydroxyaldehyde, which can be easily dehydrated:

Now try questions 4.8 and 4.9

4.2

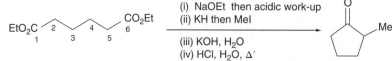

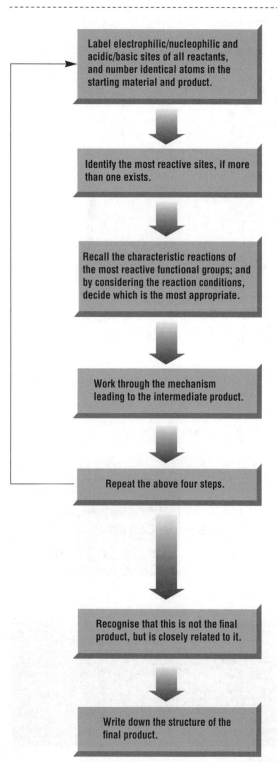

| Label electrophilic/nucleophilic and acidic/basic sites of all reactants, and number identical atoms in the starting material and product. | The carbonyl group is polarised by the *electronegativity* difference between carbon and oxygen, making the carbon electrophilic. The α-protons of a carbonyl group are acidified by *inductive* withdrawl, and the resulting enolate is resonance stabilised and a good nucleophile. NaOEt is a good base. |

C–2 and C–5 are the only acidic sites (pK$_a$ approx. 25), and sodium ethoxide is the most basic substrate.

Identify the most reactive sites, if more than one exists.

Deprotonation of the α-protons of the ester is easily achieved by NaOEt to give an enolate, which is resonance stabilised, and which is also a good nucleophile. Esters readily undergo nucleophilic *addition–elimination*.

Recall the characteristic reactions of the most reactive functional groups; and by considering the reaction conditions, decide which is the most appropriate.

Intramolecular attack of the enolate at the other carbonyl gives a ketone. One of the products (ethoxide) is a good base, capable of removing an acidic proton in the product and driving the *equilibrium* to the right.

Work through the mechanism leading to the intermediate product.

* MeI is a good electrophile, by virtue of the electronegativity difference of carbon and iodine; *nucleophilic substitution* by the ketone enolate gives the methylated product.
* KOH is a good nucleophile and base; ester hydrolysis by *addition–elimination* gives the corresponding carboxylate anion.
* Acidification and heating causes *decarboxylation*, to give the final product, in its enol form.

Repeat the above four steps.

Recognise that this is not the final product, but is closely related to it.

Tautomerisation of the enol to the keto form gives the final product.

Write down the structure of the final product.

100

Summary: This is an example of the Dieckmann cyclisation. Hydrolysis and decarboxylation of the products gives a cyclic ketone:

$$EtO_2C \diagup\!\!\diagdown (CH_2)_n \diagdown\!\!\diagup CO_2Et \xrightarrow{\text{NaOEt}} \text{(cyclic ketone)}(CH_2)_n$$

Now try questions 4.10 and 4.16

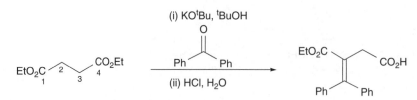

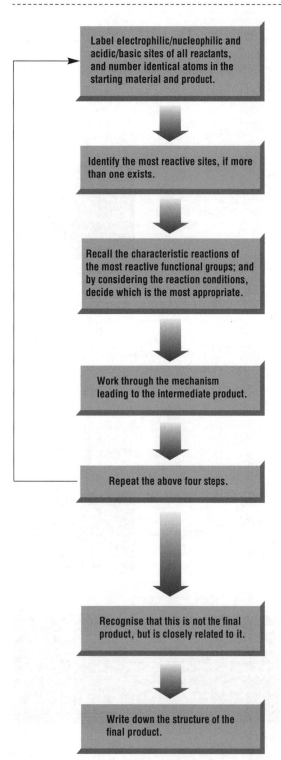

Label electrophilic/nucleophilic and acidic/basic sites of all reactants, and number identical atoms in the starting material and product.	The carbonyl group is polarised by the *electronegativity* difference between carbon and oxygen, making the carbon electrophilic. The α-protons of a carbonyl group are acidified by *inductive* withdrawl, and the resulting enolate is resonance stabilised and a good nucleophile. KO^{t}Bu is a very strong base, but a poor nucleophile, due to steric hindrance of the branched alkyl group.
Identify the most reactive sites, if more than one exists.	C-2 and C-3 are the only acidic sites in the starting material, and potassium t-butoxide is the strongest base. Benzophenone is electrophilic at the ketone carbonyl.
Recall the characteristic reactions of the most reactive functional groups; and by considering the reaction conditions, decide which is the most appropriate.	Ester enolates are good nucleophiles; ketones are good electrophiles, and readily undergo *nucleophilic addition.*
Work through the mechanism leading to the intermediate product.	*Nucleophilic addition* of the ester enolate to the ketone occurs. The alkoxide so generated will undergo a *nucleophilic-addition* reaction with the terminal ester, and a 5-membered *lactone* is formed as it is thermodynamically favoured.
Repeat the above four steps.	Strong acid protonates the lactone carbonyl oxygen, converting it to a good *leaving group*; elimination of carboxylate then occurs to give a highly stabilised *carbocation* intermediate, which is then deprotonated to give the alkene product.
Recognise that this is not the final product, but is closely related to it.	Acidification on work-up gives the carboxylic acid product.
Write down the structure of the final product.	

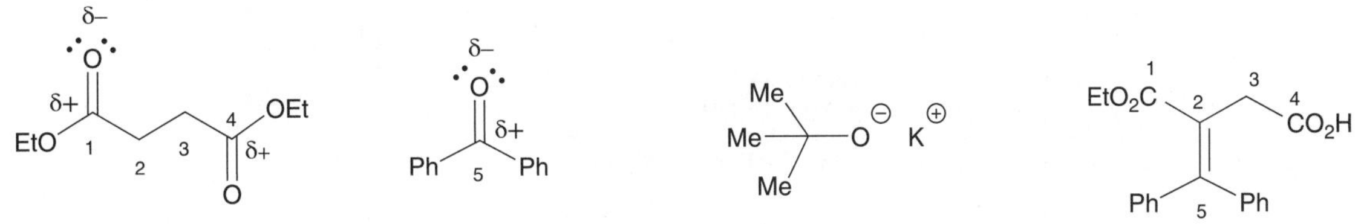

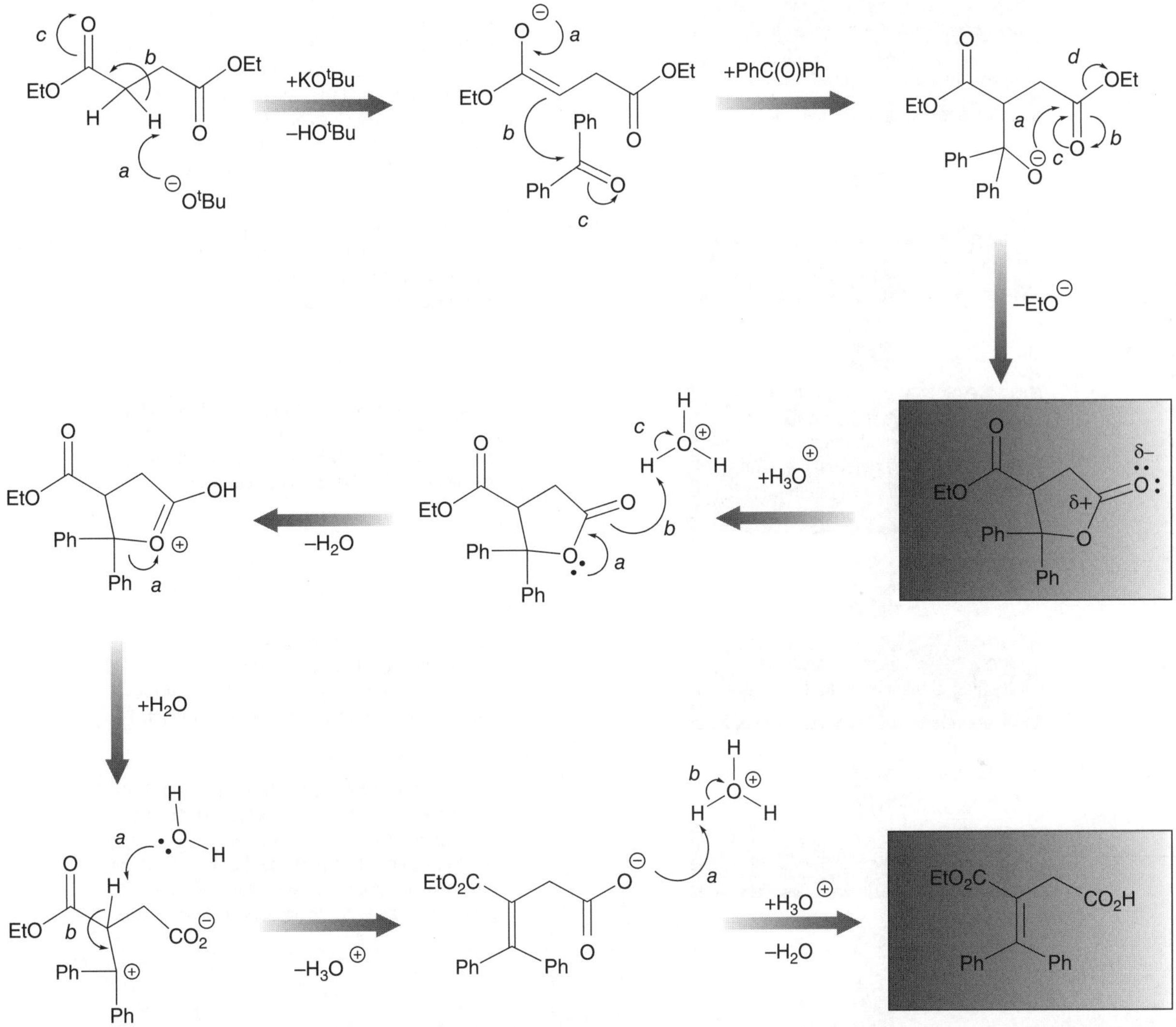

δ−
δ+ EtO 1 2 3 4 OEt δ+
O   O

δ−
Ph 5 Ph δ+
O

Me
Me—O⊖ K⊕
Me

EtO₂C 1 2 3 4 CO₂H
Ph 5 Ph

- -

c
EtO b OEt
H H
a ⊖O^tBu
→ +KO^tBu / −HO^tBu →

O⊖ a
EtO OEt
b
Ph
Ph O c
→ +PhC(O)Ph →

O
EtO d OEt
a
Ph b
Ph O⊖ c
→ −EtO⊖ →

O
EtO O
δ+
Ph O δ−
Ph

← +H₃O⊕

O
EtO O
c H—O⊕—H
H
b
Ph O a
Ph

← −H₂O

O
EtO OH
Ph O⊕
Ph a

↓ +H₂O

O a H—O—H
EtO H
b
Ph ⊕ CO₂⊖
Ph

→ −H₃O⊕ →

EtO₂C
O⊖
Ph Ph O
b H—O⊕—H
H
a

→ +H₃O⊕ / −H₂O →

EtO₂C CO₂H
Ph Ph

Summary: This is an example of the Stobbe reaction; noteworthy in this reaction is that the condensation proceeds with concomitant hydrolysis of the terminal ester:

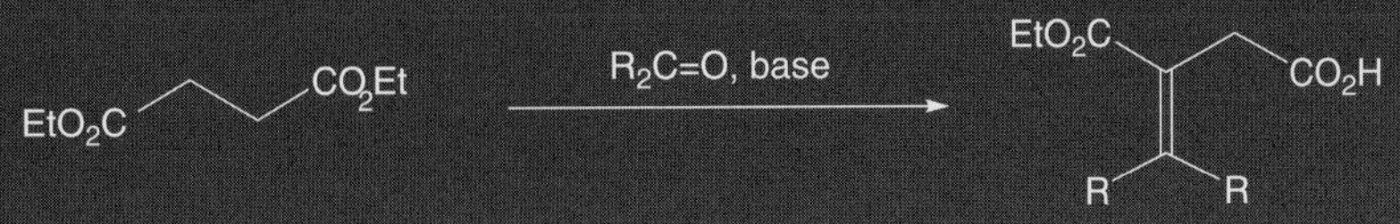

EtO₂C CO₂Et
→ R₂C=O, base →
EtO₂C CO₂H
R R

Now try questions 4.11 and 4.17

4.4

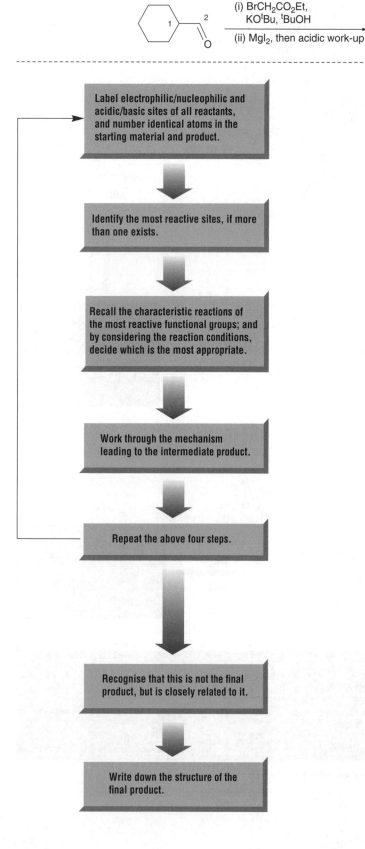

The carbonyl group is polarised by the *electronegativity* difference between carbon and oxygen, making the carbon electrophilic. In a bromoester, the α-protons of a carbonyl group are acidified by *inductive* withdrawl both from the carbonyl group and the halogen, and the resulting enolate is resonance stabilised and a good nucleophile. KO^tBu is a very strong base, but a poor nucleophile, due to steric hindrance.

The aldehyde is the only electrophilic site, and the most acidic hydrogens are located between the carbonyl and halogen groups. Potassium *t*-butoxide is the strongest base.

Aldehydes are susceptible to *nucleophilic addition*. The α-hydrogens of a bromoester are especially acidic, and can easily be removed by *t*-butoxide, giving an enolate which is nucleophilic.

Nucleophilic addition to the aldehyde gives an alkoxide, which readily undergoes an intramolecular *nucleophilic substitution* at the adjacent bromide to give an *epoxide* product.

MgI$_2$ is a good Lewis acid, which will co-ordinate to the electron rich epoxide oxygen, and catalyse *nucleophilic substitution* of the epoxide by iodide; the regioselectivity is given by the stability of the intermediate secondary carbocation which is generated if an S$_N$1-like mechanism is assumed.

Acidic work-up protonates the alkoxide, to give the product.

Label electrophilic/nucleophilic and acidic/basic sites of all reactants, and number identical atoms in the starting material and product.

Identify the most reactive sites, if more than one exists.

Recall the characteristic reactions of the most reactive functional groups; and by considering the reaction conditions, decide which is the most appropriate.

Work through the mechanism leading to the intermediate product.

Repeat the above four steps.

Recognise that this is not the final product, but is closely related to it.

Write down the structure of the final product.

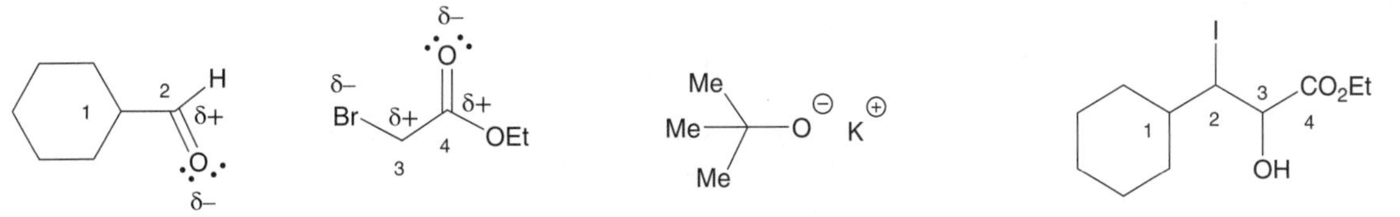

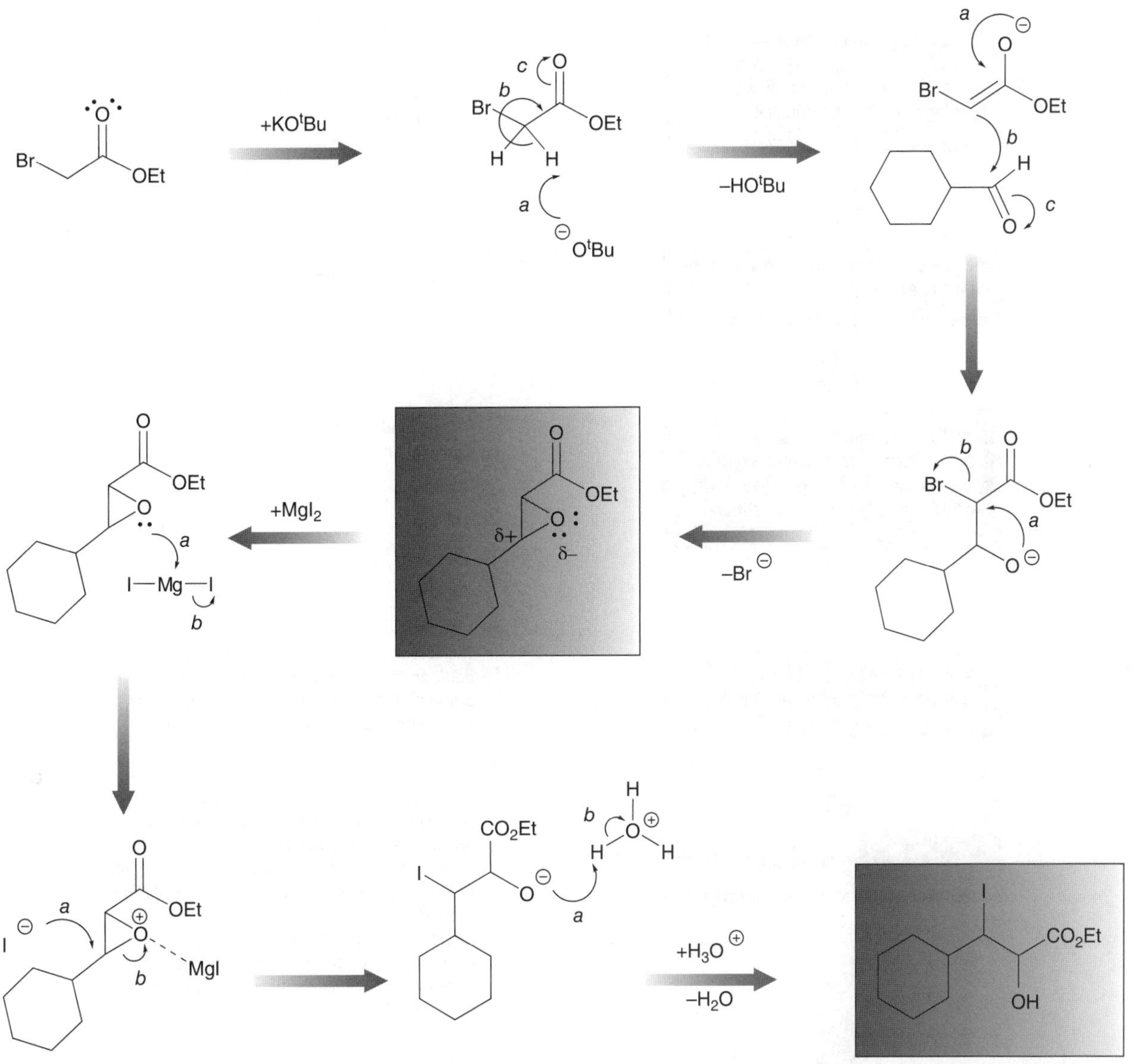

+KO^{t}Bu

−HO^{t}Bu

−Br$^{\ominus}$

+MgI$_2$

+H$_3$O$^{\oplus}$
−H$_2$O

OEt

OtBu

CO$_2$Et

OH

I—Mg—I

I$^{\ominus}$

MgI

Summary: This is an example of the Darzens reaction:

XCH$_2$CO$_2$Et + R$_2$C=O $\xrightarrow{\text{base}}$ R$_2$C(—O—)CO$_2$Et

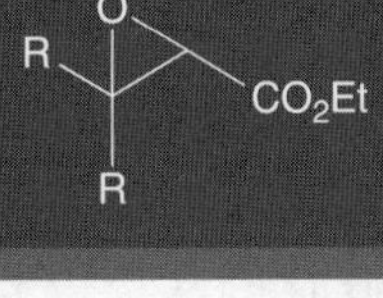

X = Br, Cl

Now try questions 4.12 and 4.18

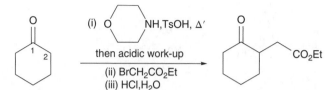

Label electrophilic/nucleophilic and acidic/basic sites of all reactants, and number identical atoms in the starting material and product.

The carbonyl group is polarised by the *electronegativity* difference between carbon and oxygen, making the carbon electrophilic. The α-protons of a carbonyl group are acidified by *inductive* withdrawl, and the resulting enolate is resonance stabilised and a good nucleophile. However, although morpholine is a good nucleophile, it is not basic enough to deprotonate a ketone.

Identify the most reactive sites, if more than one exists.

The ketone is the only electrophilic site, and morpholine the most nucleophilic reagent.

Recall the characteristic reactions of the most reactive functional groups; and by considering the reaction conditions, decide which is the most appropriate.

Ketones and secondary amines readily react to give enamines.

Work through the mechanism leading to the intermediate product.

The ketone is protonated, and *nucleophilic addition* of morpholine to the ketone then proceeds, but in this case only *elimination* of the intermediate to an *enamine* is possible.

Repeat the above four steps.

* An enamine is an excellent nucleophile (compare with an enol) at the β-carbon since nitrogen is strongly nucleophilic. Ethyl bromoacetate is a good electrophile, with a good Br⁻ leaving group, and *nucleophilic substitution* then occurs.
* Acid catalysed iminium ion hydrolysis (compare hydrolysis of an acetal) then occurs; ester hydrolysis does not occur under these mild conditions.

Recognise that this is not the final product, but is closely related to it.

Not needed here.

Write down the structure of the final product.

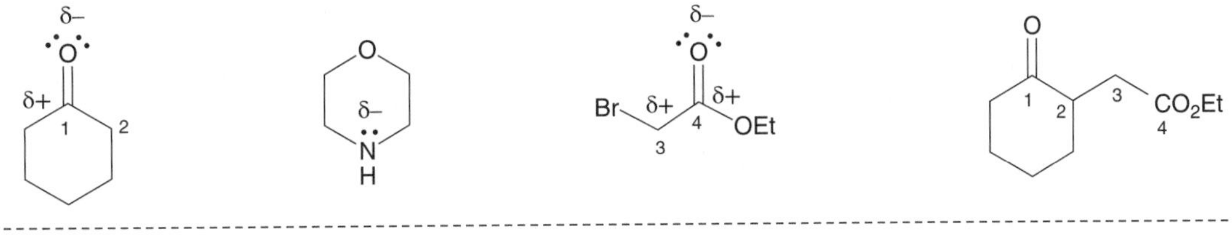

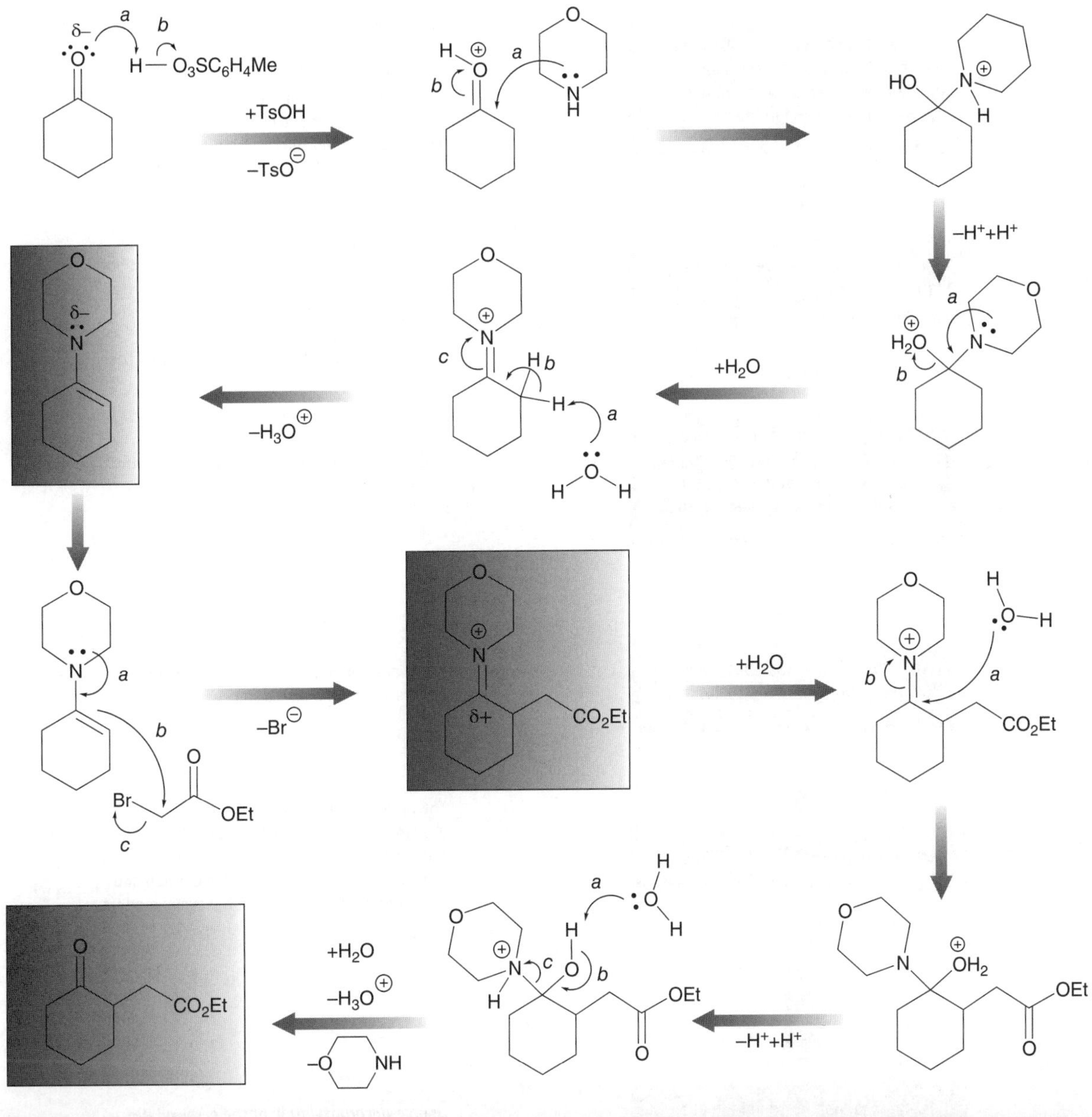

δ−

O

δ+ 1 2

O

δ− N—H

δ− O

Br δ+ 3 δ+ 4 OEt

O

1 2 3 CO₂Et 4

---

a

b

δ− H—O₃SC₆H₄Me

+TsOH

−TsO⁻

H⊕ O

b a

O N—H

HO N⊕ H

−H⁺+H⁺

O N

δ−

O N⊕

c Hb

H a

H O H

+H₂O

a O N⊕

H₂O b

+H₂O

O N

b a H⁺ O H CO₂Et

O N⊕ δ+ CO₂Et

+H₂O

O N⊕ OH₂ OEt O

O N

a

b

Br O OEt

c

−Br⁻

O N

b c O OEt

−H⁺+H⁺

O

CO₂Et

+H₂O

−H₃O⊕

−O NH

a H O H

H O b

O N⊕ H c

OEt

O

Summary: This is an example of the Stork enamine synthesis; this reaction is equivalent to the alkylation of an enolate.:

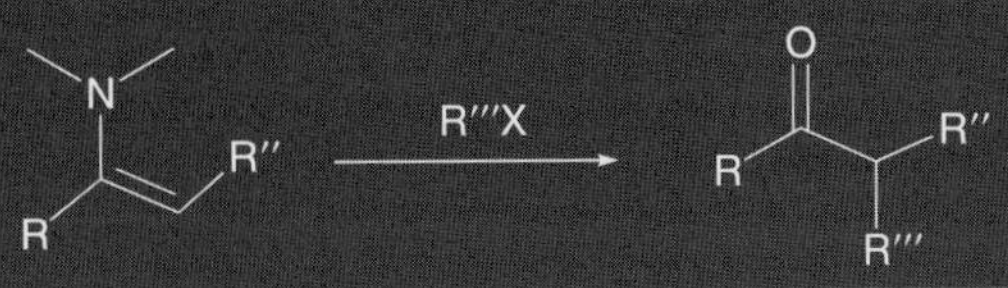

N

R R''

R'''X

O

R R'' R'''

Now try questions 4.13 and 4.19

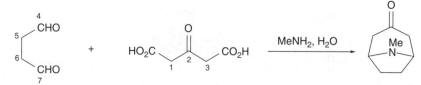

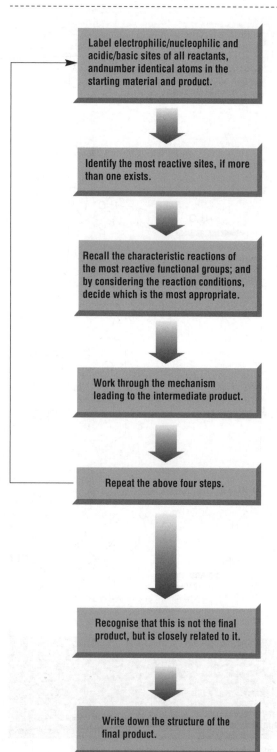

Aldehydes are good electrophiles, since the carbonyl group is polarised by the electronegativity difference between carbon and oxygen. 1,3-Dicarbonyl compounds are highly acidic at the α-position (here labelled C-1 and C-3) and the resulting enol or enolate is a good nucleophile. Methylamine is a good base and nucleophile.

The dialdehyde is a very reactive electrophile, and susceptible to *nucleophilic addition* at both C4 and C-7. Acetone dicarboxylic acid exists in equilibrium with its enolic *tautomer*, and is a therefore a good nucleophile at C-3. Dimethylamine is a good nucleophile and base.

Aldehydes readily react with amines to form *imines* with elimination of water.

Addition-elimination reaction of the amine generates an *iminium* cation.

The *iminium* cation, which is highly electrophilic, is easily attacked by the nucleophilic enol; this whole process is repeated using the other aldehyde group and the alternative enol, to form a bicyclic product.

Double *decarboxylation* of the carboxylate anion gives the product.

Summary: This is an example of the Mannich reaction:

Now try questions 4.14 and 4.20

4.7

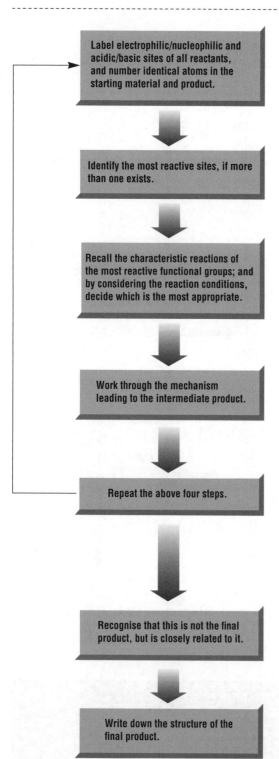

The α-protons of β-dicarbonyl compounds are highly acidic (pK_a about 13) and give an enolate which is a good nucleophile. α, β-Unsaturated ketones are electrophilic at C–2, and at C–4 by conjugation. EtO⁻ is a good base.

α, β-Unsaturated ketones are electrophilic at the β-position (here labelled C–4). The a-position of the β-dicarbonyl compound is the most acidic, and the corresponding enolate is the most nucleophilic (here labelled C–6). Sodium ethoxide is the strongest available base,

α, β-Unsaturated carbonyl compounds readily undergo 1,4-nucleophilic addition reactions with enolate nucleophiles.

Conjugate addition of the enolate to the *Michael acceptor*, followed by *equilibration* of the ketone enolate to the alternative (less stable) enolate; ketone formation then follows by standard *addition-elimination* mechanism at an ester. The overall reaction is under thermodynamic control, and proceeds because of the irreversible ring closure step.

Acid catalysed *hydrolysis* of the ester, followed by standard *decarboxylation* of the resulting β-ketoacid, gives and enol product.

This intermediate product can be converted to the final product by *tautomerisation*, a process that can be catalysed by any available acid or base.

Flowchart:

- Label electrophilic/nucleophilic and acidic/basic sites of all reactants, and number identical atoms in the starting material and product.
- Identify the most reactive sites, if more than one exists.
- Recall the characteristic reactions of the most reactive functional groups; and by considering the reaction conditions, decide which is the most appropriate.
- Work through the mechanism leading to the intermediate product.
- Repeat the above four steps.
- Recognise that this is not the final product, but is closely related to it.
- Write down the structure of the final product.

Now try questions 4.15, 4.21 and 4.22

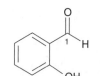 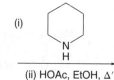 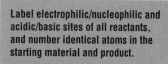

(i)

(ii) HOAc, EtOH, Δ′

EtO₂C

CO₂Et

Label electrophilic/nucleophilic and acidic/basic sites of all reactants, and number identical atoms in the starting material and product.

The carbonyl group is polarised by the *electronegativity* difference between carbon and oxygen, making the carbon of the aldehyde electrophilic. The α-protons of a carbonyl group are acidified by *inductive* withdrawl, and the resulting enolate is resonance stabilised and a good nucleophile. Piperidine (pK$_a$ 11.2) is a good base, and able to catalyse formation of the enolic form of malonate.

The aldehyde is the most reactive electrophilic site, diethyl malonate the most acidic reagent (pKa 13) and morpholine the most basic reagent.

Identify the most reactive sites, if more than one exists.

Recall the characteristic reactions of the most reactive functional groups; and by considering the reaction conditions, decide which is the most appropriate.

Stabilised enolates readily participate with reactive aldehydes and ketones in the aldol reaction.

Work through the mechanism leading to the intermediate product.

Enolisation of the diester is followed by standard *nucleophilic addition* to the aldehyde carbonyl group of the starting material. This is followed by E$_1$CB elimination of water to give the α, β-unsaturated product.

Repeat the above four steps.

Treatment with acid activates the ester carbonyl group to addition-elimination by the adjacent hydroxy group, leading to ring closure and cyclisation to give the 6-membered lactone.

Recognise that this is not the final product, but is closely related to it.

Not needed here.

Write down the structure of the final product.

Now try question 4.9

4.9

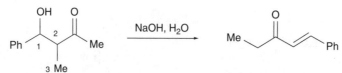

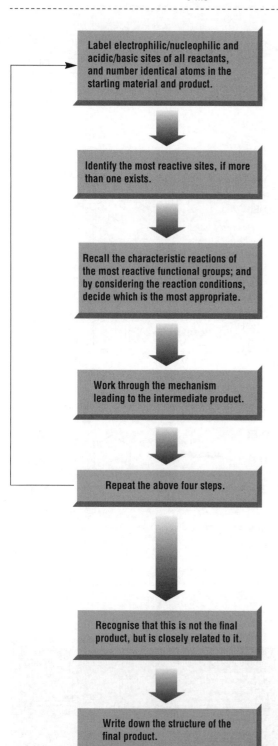

Label electrophilic/nucleophilic and acidic/basic sites of all reactants, and number identical atoms in the starting material and product.	The carbonyl group is polarised by the *electronegativity* difference between carbon and oxygen, making the carbon electrophilic. The α-protons of a carbonyl group are also acidified by *inductive* withdrawl, and the resulting enolate is resonance stabilised and a good nucleophile. The alcohol has an acidic hydrogen. Hydroxide is an excellent base.
Identify the most reactive sites, if more than one exists.	The alcoholic O–H group has the most acidic hydrogen, and is readily deprotonated under these basic conditions.
Recall the characteristic reactions of the most reactive functional groups; and by considering the reaction conditions, decide which is the most appropriate.	Stabilised enolates readily participate with reactive aldehydes and ketones in the aldol reaction; this reaction may also run in the reverse direction, as is relevant here.
Work through the mechanism leading to the intermediate product.	
Repeat the above four steps.	Deprotonation of the hydroxide group generates an alkoxide which collapses in a reverse aldol process to generate benzaldehyde and an enolate. This enolate may equilibrate to form another (less stable) enolate, which re-adds to benzaldehyde to make a new aldol adduct. This in turn eliminates water to form a new α, β-unsaturated ketone product. The whole process is driven by the formation of the most stable product (*thermodynamic control*).
Recognise that this is not the final product, but is closely related to it.	Not needed here.
Write down the structure of the final product.	

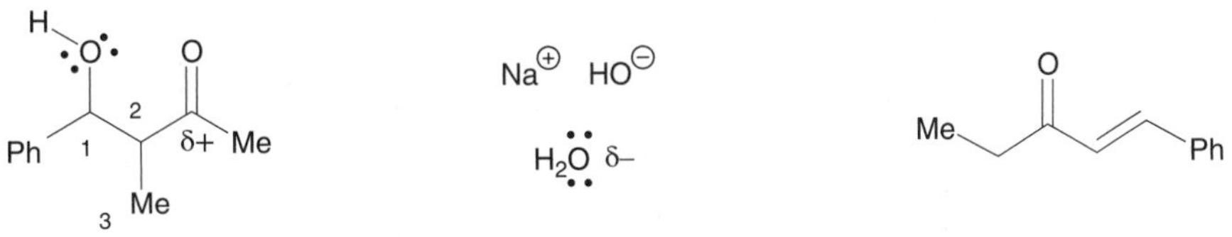

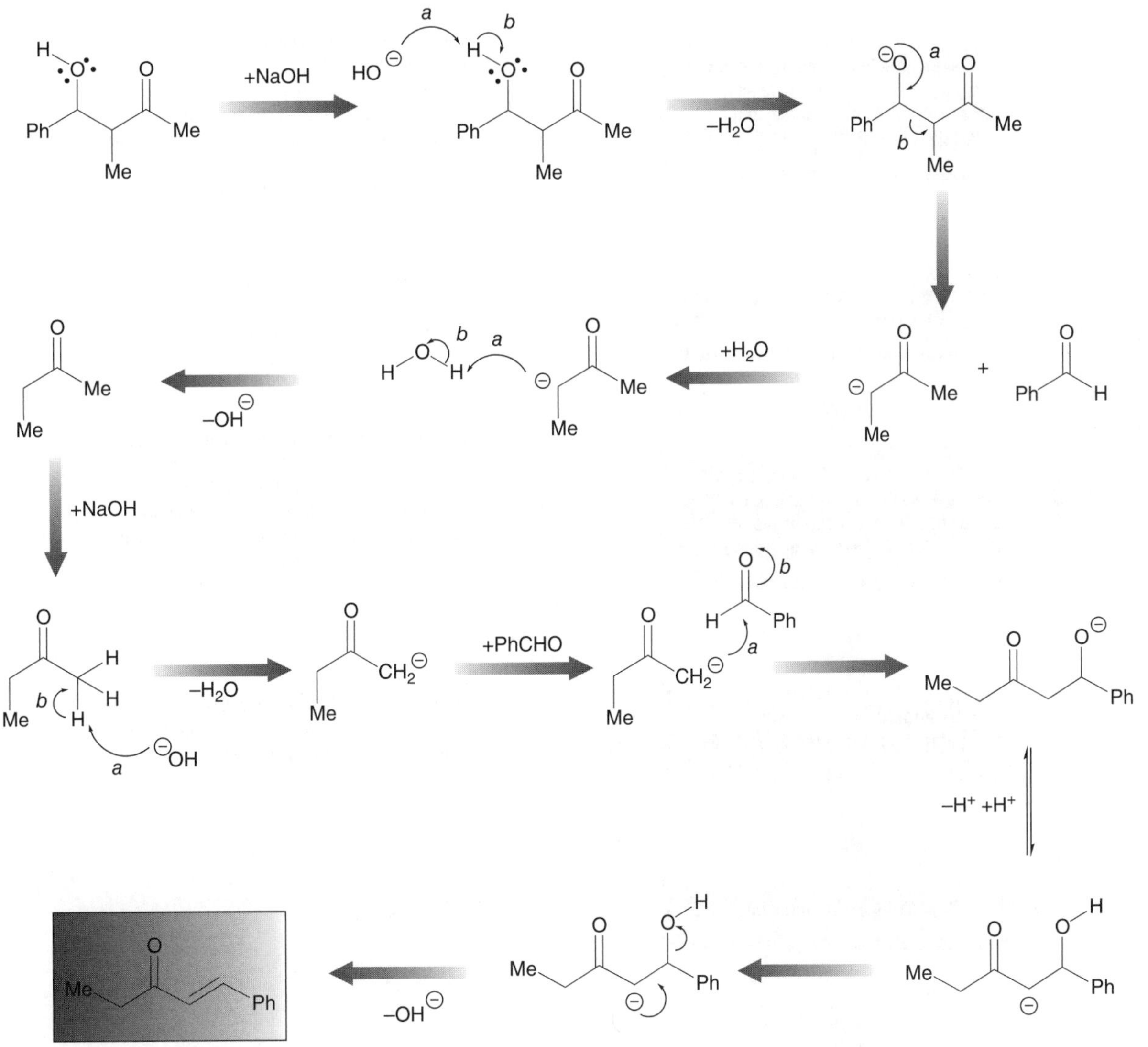

$Na^{\oplus}$ $HO^{\ominus}$

H_2O $\delta-$

Me ... O ... Ph

+NaOH

$HO^{\ominus}$ a b H

$-H_2O$

$\ominus$O a b Me

Ph Me Me

$\ominus$ O Me + Ph O H

Me

+H_2O

H O b a H $\ominus$ O Me

Me

$-OH^{\ominus}$

O Me

Me

+NaOH

O H H b Me H H a $OH^{\ominus}$

$-H_2O$

O $CH_2^{\ominus}$ Me

+PhCHO

O H b Ph a $CH_2^{\ominus}$ Me

Me O $O^{\ominus}$ Ph

$-H^+$ $+H^+$

Me O O H Ph $\ominus$

Me O O H Ph $\ominus$

$-OH^{\ominus}$

Me O Ph

Summary: This is an example of the reverse of an aldol reaction, which results in the formation of two aldehydes or ketones from give a β–hydroxyaldehyde:

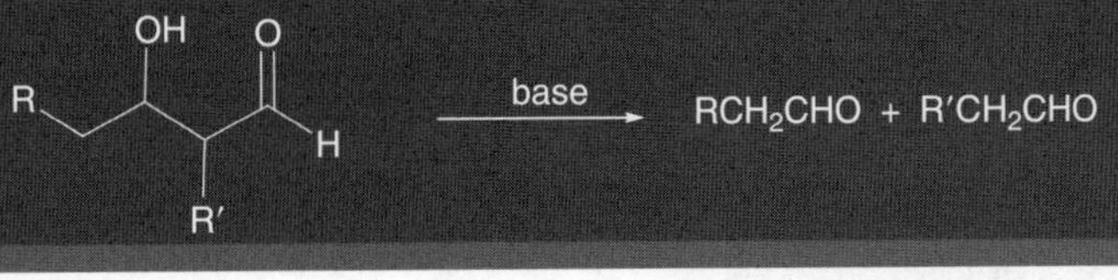

OH O

R R′ H

$\xrightarrow{\text{base}}$ RCH_2CHO + $R'CH_2CHO$

115

4.10

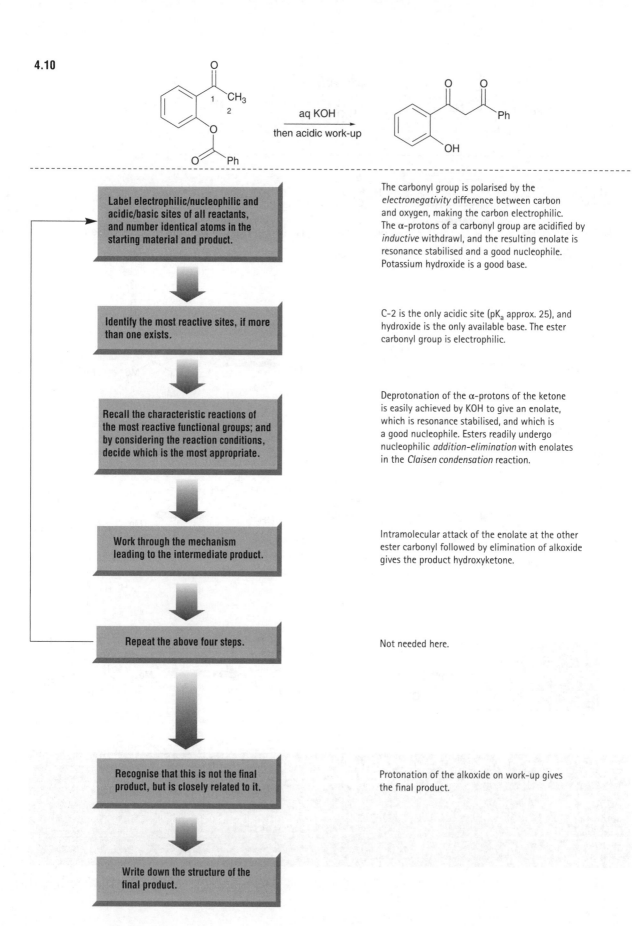

Label electrophilic/nucleophilic and acidic/basic sites of all reactants, and number identical atoms in the starting material and product.	The carbonyl group is polarised by the *electronegativity* difference between carbon and oxygen, making the carbon electrophilic. The α-protons of a carbonyl group are acidified by *inductive* withdrawl, and the resulting enolate is resonance stabilised and a good nucleophile. Potassium hydroxide is a good base.
Identify the most reactive sites, if more than one exists.	C-2 is the only acidic site (pK$_a$ approx. 25), and hydroxide is the only available base. The ester carbonyl group is electrophilic.
Recall the characteristic reactions of the most reactive functional groups; and by considering the reaction conditions, decide which is the most appropriate.	Deprotonation of the α-protons of the ketone is easily achieved by KOH to give an enolate, which is resonance stabilised, and which is a good nucleophile. Esters readily undergo nucleophilic *addition-elimination* with enolates in the *Claisen condensation* reaction.
Work through the mechanism leading to the intermediate product.	Intramolecular attack of the enolate at the other ester carbonyl followed by elimination of alkoxide gives the product hydroxyketone.
Repeat the above four steps.	Not needed here.
Recognise that this is not the final product, but is closely related to it.	Protonation of the alkoxide on work-up gives the final product.
Write down the structure of the final product.	

Summary: This is an example of the Claisen reaction, in which two esters are condensed to give a β-dicarbonyl product:

$$RCH_2CO_2Et + R'CH_2CO_2Et \xrightarrow{\text{base}}$$

Now try question 4.16

4.11

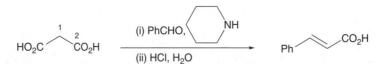

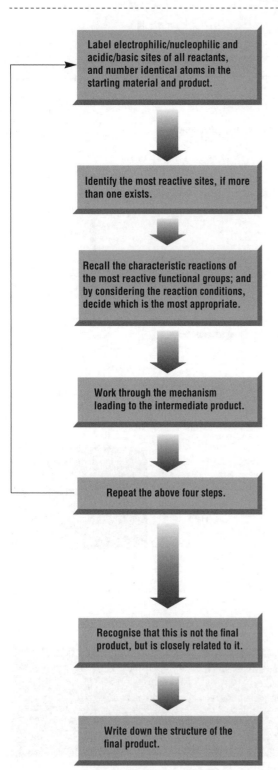

The carbonyl group is polarised by the *electronegativity* difference between carbon and oxygen, making the carbon of the aldehyde electrophilic. The α-protons of a carbonyl group are acidified by *inductive* withdrawl, and the resulting enolate is resonance stabilised and a good nucleophile; malonic acid is especially activated in this way. Piperidine is a weak base, but also a good nucleophile.

The aldehyde is the most electrophilic site, and C-1 and the carboxylic acid groups are both acidic sites. Piperidine is the only base and a good nucleophile.

β-Dicarbonyl groups readily react as nucleophiles through the α-centre with a variety of electrophiles.

The highly reactive aldehyde forms an iminium ion with the benzaldehyde, which is a particularly reactive electrophile. Initial deprotonation of the carboxylic acid permits equilibration to the malonyl enolate by removal of an α-proton, which is resonance stabilised, and which is a good nucleophile. These react together in an aldol-like reaction, to give β-aminoacid. Proton equilibration generates a good leaving group which can be expelled by loss of CO_2, to give the product.

Not needed here.

118

Summary: This is an example of the Knoevenagel reaction:

Now try question 4.17

4.12

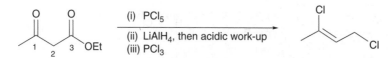

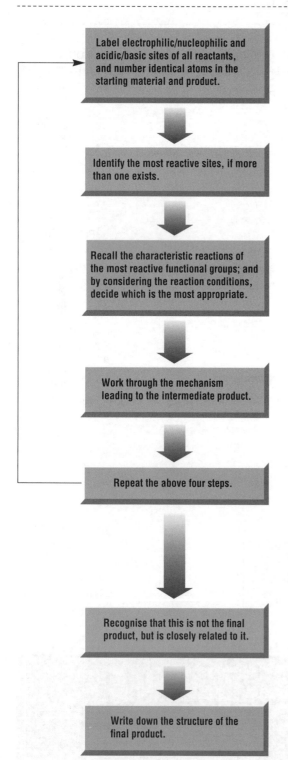

Label electrophilic/nucleophilic and acidic/basic sites of all reactants, and number identical atoms in the starting material and product.	The α-protons of the β-dicarbonyl group are acidified by *inductive* withdrawl, and the resulting enol is resonance stabilised and a good nucleophile. Phosphorus pentachloride is an excellent electrophile.
Identify the most reactive sites, if more than one exists.	C-2 is the only acidic site (pK$_a$ approx. 25) and the keto form is in equilibrium with its enol form.
Recall the characteristic reactions of the most reactive functional groups; and by considering the reaction conditions, decide which is the most appropriate.	The enol readily reacts with suitable electrophiles, in this case phosphorus pentachloride, by *nucleophilic substitution*.
Work through the mechanism leading to the intermediate product.	Attack by the enol at phosphorus gives an activated enol, which may further react by a nucleophilic *addition-elimination* process with chloride, along with formation of a P=O containing product, which provides the driving force for the reaction.
Repeat the above four steps.	* The ester product is readily reduced by LiAlH$_4$, by two successive additions of nucleophilic addition of hydride anion (H$^-$), to give an alkoxide product. * Reaction of the alcohol with phosphorus trichloride generates an activated intermediate which is displaced with chloride to give the final product, an allyl chloride.
Recognise that this is not the final product, but is closely related to it.	Not needed here.
Write down the structure of the final product.	

Summary: This is an example of the nucleophilicity of β-dicarbonyl groups, which may react via the enolic oxygen:

Now try question 4.18

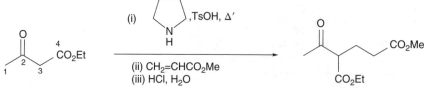

Label electrophilic/nucleophilic and acidic/basic sites of all reactants, and number identical atoms in the starting material and product.

The carbonyl group is polarised by the *electronegativity* difference between carbon and oxygen, making the carbon electrophilic. The α-protons of a carbonyl group are acidified by *inductive* withdrawl, and the resulting enolate is resonance stabilised and a good nucleophile. Pyrrolidine is a very good nucleophile, but not basic enough to deprotonate a ketone.

Identify the most reactive sites, if more than one exists.

The ketone is the most reactive electrophilic site, being more reactive than the ester, and pyrrolidine the most nucleophilic reagent.

Recall the characteristic reactions of the most reactive functional groups; and by considering the reaction conditions, decide which is the most appropriate.

Ketones and amines readily react to give imines; but in the case of secondary amines, the first formed iminium species is converted to the enamine.

Work through the mechanism leading to the intermediate product.

The ketone is protonated, and *nucleophilic addition* of pyrrolidine to the ketone then proceeds. In this case *elimination* of the first formed iminium intermediate to give an *enamine* is possible.

Repeat the above four steps.

* An enamine is an excellent nucleophile (compare with an enol) at the β-carbon since nitrogen is strongly nucleophilic. Methyl acrylate is a good electrophile, but is most reactive at the β-position, and *conjugate or Michael addition* then occurs. This gives an intermediated iminium ion.
* Acid catalysed iminium ion hydrolysis (compare hydrolysis of an acetal) gives the ketone product; ester hydrolysis does not occur under these mild conditions.

Recognise that this is not the final product, but is closely related to it.

Not needed here.

Write down the structure of the final product.

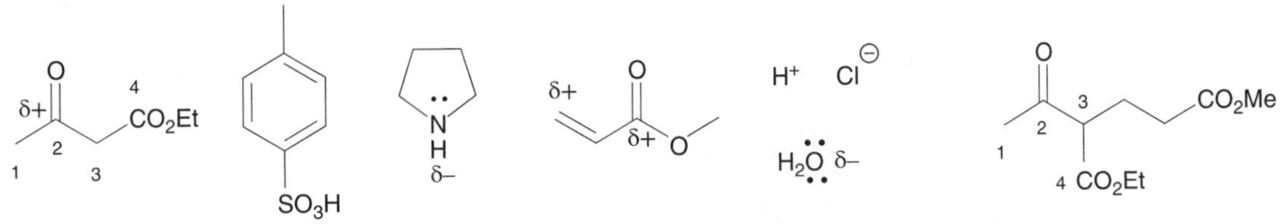

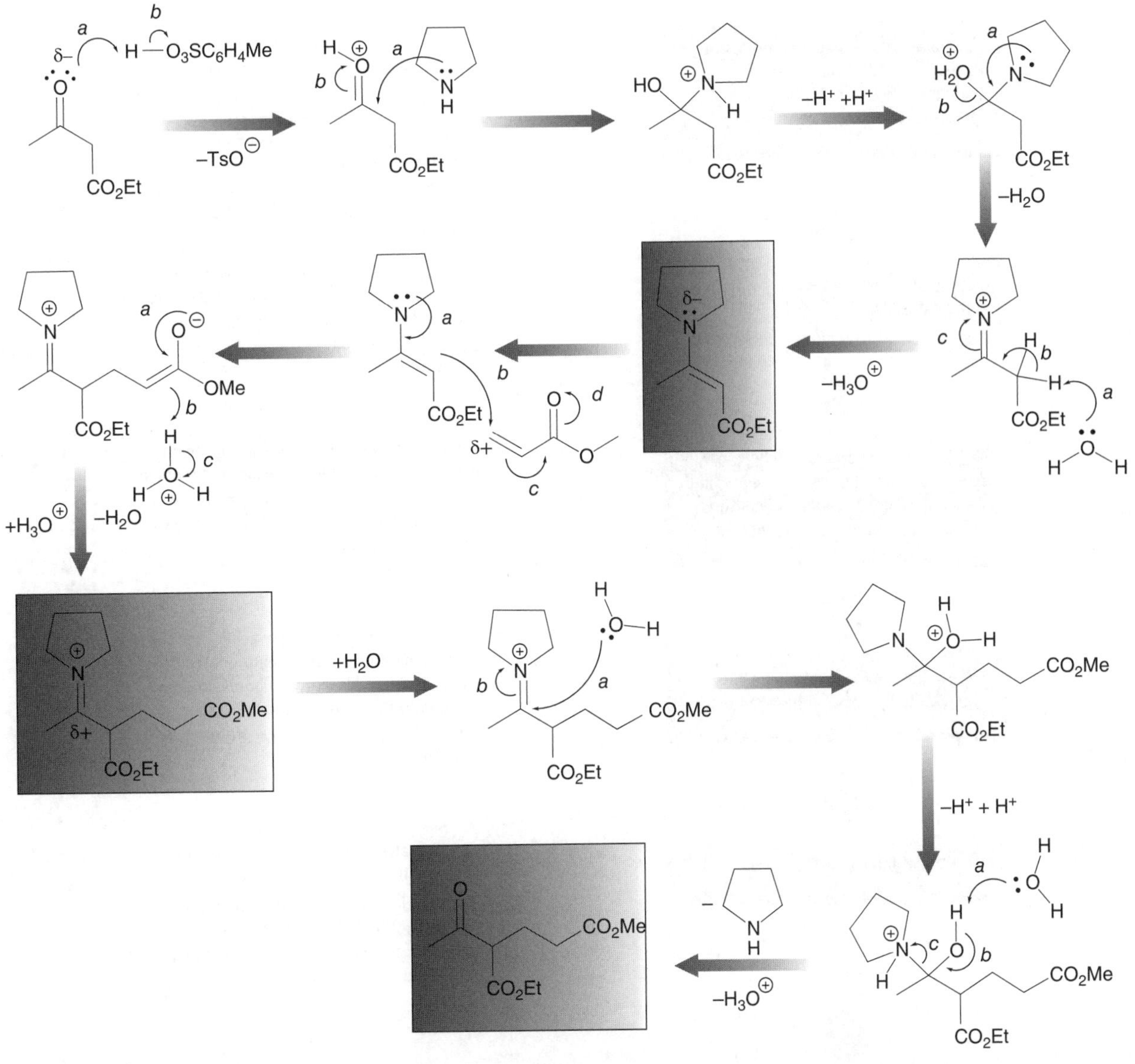

Summary: This is an example of alkylations of enamines using a Michael acceptor (methyl acrylate) as electrophile; it is called the Stork enamine synthesis.:

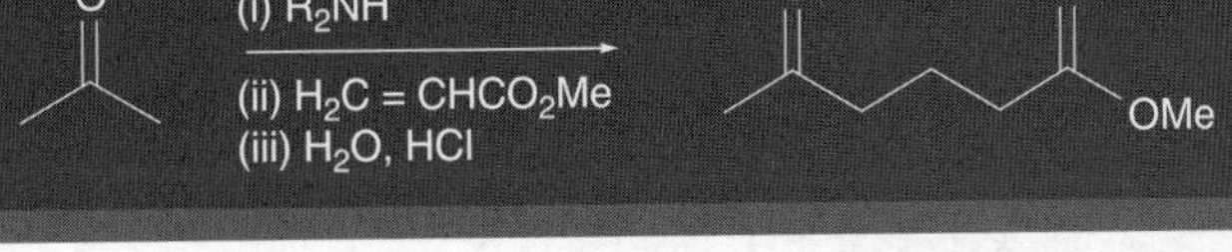

(i) R_2NH
(ii) $H_2C = CHCO_2Me$
(iii) H_2O, HCl

Now try question 4.19

123

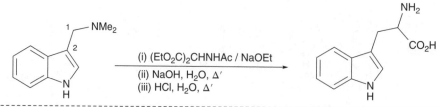

Label electrophilic/nucleophilic and acidic/basic sites of all reactants, and number identical atoms in the starting material and product.

The carbonyl group is polarised by the *electronegativity* difference between carbon and oxygen, making the carbon electrophilic. The α-protons of a carbonyl group are acidified by *inductive* withdrawl, and the resulting enolate is resonance stabilised and a good nucleophile. Gramine is a precursor of an α, β-unsaturated system, by elimination of dimethylamine, which is a good electrophile. Sodium ethoxide is an excellent base.

Identify the most reactive sites, if more than one exists.

The β-dicarbonyl group is the most acidic site, being easily deprotonated by ethoxide under the conditions of the reaction, and the α, β-unsaturated imine is the most electrophilic site.

Recall the characteristic reactions of the most reactive functional groups; and by considering the reaction conditions, decide which is the most appropriate.

β-Dicarbonyl groups are very acidic, being easily deprotonated by ethoxide, and forming a very reactive nucleophilic enolate. This will react with a suitable electrophile; in this case the α, β-unsaturated imine.

Work through the mechanism leading to the intermediate product.

β-Dicarbonyl groups are very acidic, and the starting material is easily deprotonated by ethoxide. This adds in a *conjugate addition* reaction to the α, β-unsaturated ketone, to give a dicarbonyl product, which is protonated on work-up.

Repeat the above four steps.

* Base catalysed ester hydrolysis gives the sodium salt of the β-oxoacid product.
* Under acidic conditions, the carboxylate anion is protonated and the resulting carboxylic acid is very prone to decarboxylation, to give the enol form of the product.

Recognise that this is not the final product, but is closely related to it.

Tautomerisation of the enol form gives the final product.

Write down the structure of the final product.

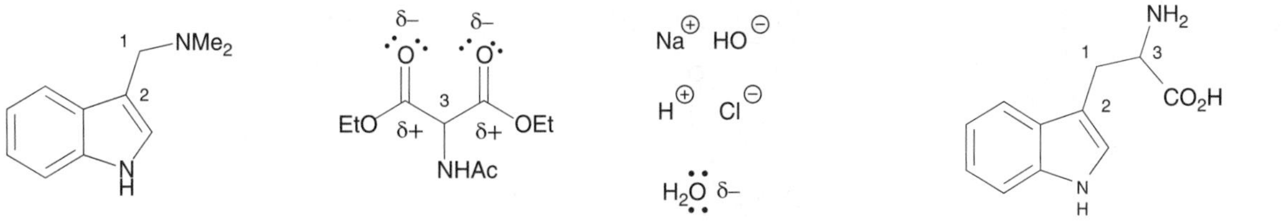

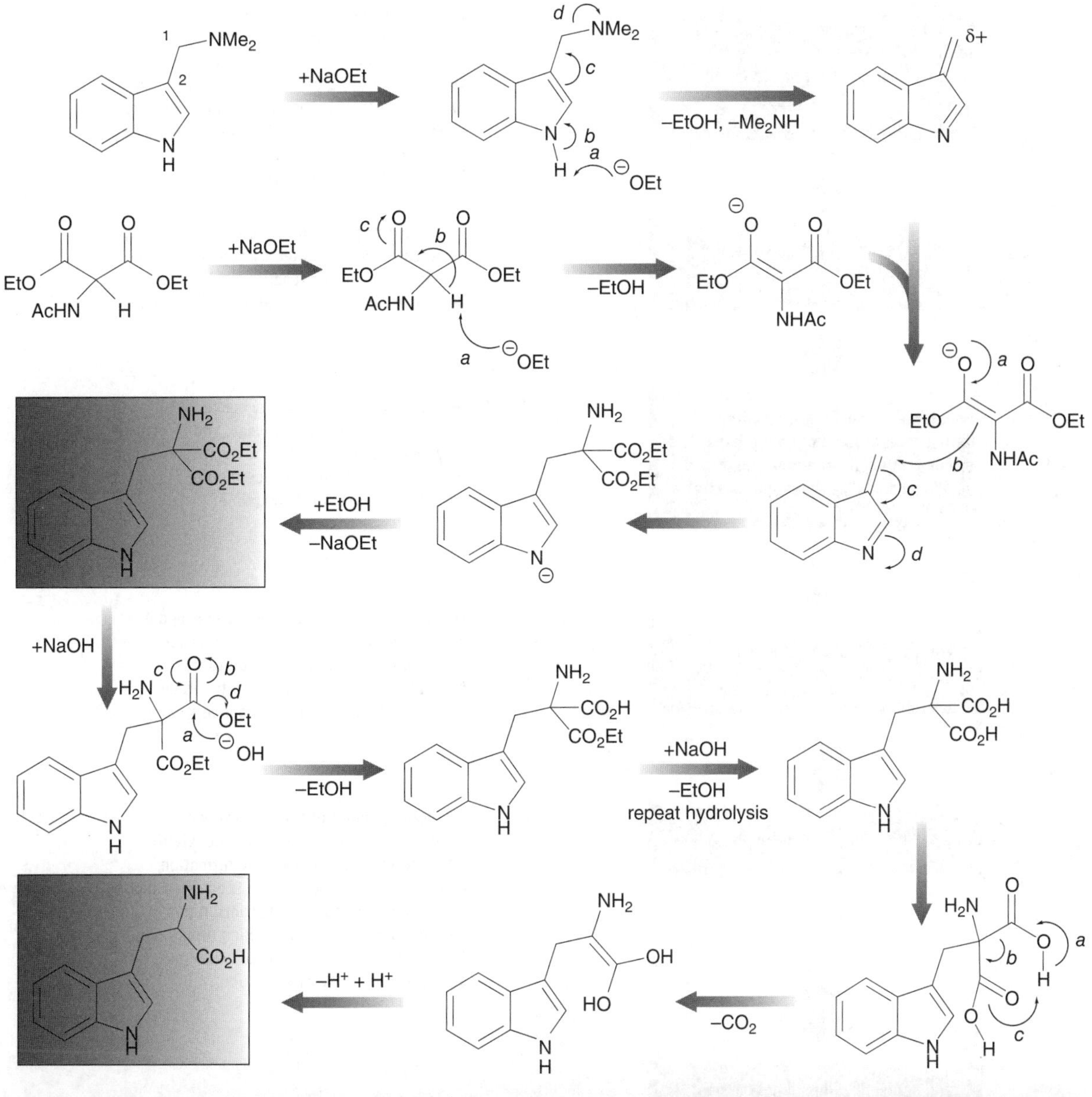

1 NMe₂
2
N
H

δ−   δ−
O   O
EtO   3   OEt
δ+   δ+
NHAc

Na⊕ HO⊖

H⊕ Cl⊖

H₂O δ−

NH₂
1   3
2   CO₂H
N
H

---

1   NMe₂
2
N   +NaOEt
H

d NMe₂
c
N   b
H   a   ⊖OEt

−EtOH, −Me₂NH

δ+
N

O   O
EtO   OEt   +NaOEt
AcHN   H

c   O   b   O
EtO   OEt
AcHN   H
a   ⊖OEt

−EtOH

⊖O   O
EtO   OEt
NHAc

⊖O   a   O
EtO   OEt
b   NHAc

c
N   d

NH₂
CO₂Et
CO₂Et
N
⊖

NH₂
CO₂Et
CO₂Et
N
H

+EtOH
−NaOEt

NH₂
—CO₂Et
—CO₂Et
N
H

+NaOH

H₂N   c   O   b
d   OEt
a   ⊖OH
CO₂Et
N
H

−EtOH

NH₂
CO₂H
CO₂Et
N
H

+NaOH
−EtOH
repeat hydrolysis

NH₂
CO₂H
CO₂H
N
H

H₂N   O
b   O   H   a
O   H
O   H   c
N
H

−CO₂

NH₂
OH
HO
N
H

−H⁺ + H⁺

NH₂
CO₂H
N
H

Summary: This is an example of enolate addition using a Michael acceptor as electrophile:

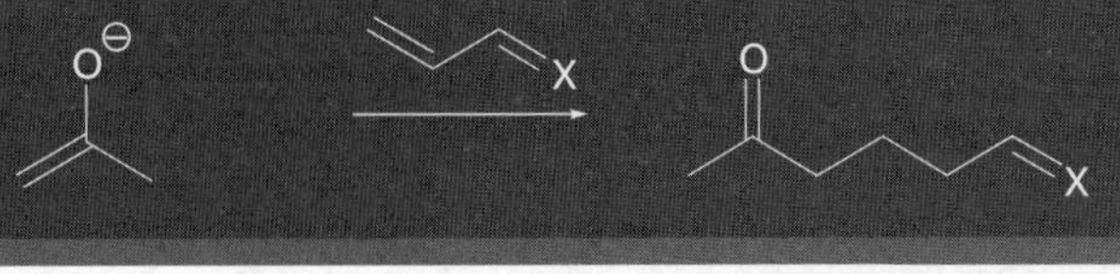

O⊖

X

O

X

Now try question 4.20

125

4.15

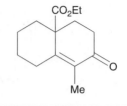

(i) NaOEt (cat.), EtOH

(ii) HCl, H₂O

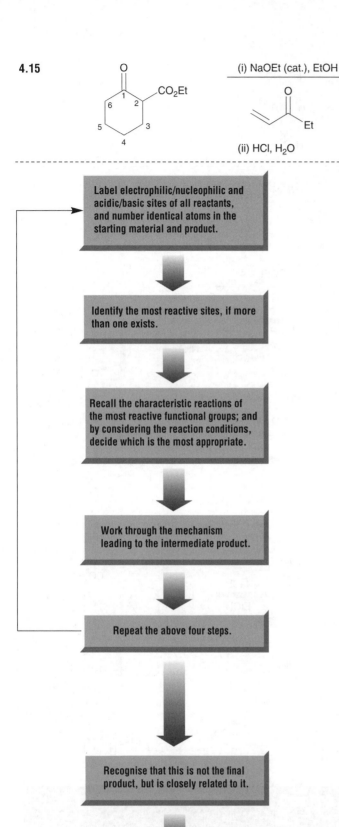

The carbonyl group is polarised by the *electronegativity* difference between carbon and oxygen, making the carbon electrophilic. The α-protons of a carbonyl group are acidified by *inductive* withdrawl, and the resulting enolate is resonance stabilised and a good nucleophile. Ethyl vinyl ketone is a good electrophile. Sodium ethoxide is an excellent base.

The β-dicarbonyl group is the most acidic site, being easily deprotonated by ethoxide under the conditions of the reaction, and the α, β-unsaturated ketone is the most electrophilic reagent.

β-Dicarbonyl groups are very acidic, being easily deprotonated by ethoxide, and forming a very reactive enolate which is nucleophilic, and will react with a suitable electrophile, in this case the α, β-unsaturated ketone.

The β-dicarbonyl group is very acidic, and is easily deprotonated by ethoxide. This adds in a *conjugate addition* reaction to the α, β-unsaturated ketone, to give a tricarbonyl product, which is protonated on work-up.

* Under the conditions of the reaction, further condensation of the product via the intermediate enol leads to ring formation, and elimination of water, in a sequence which is driven by thermodynamic control; this is an example of the Robinson Ring Annulation.
* Acid catalysed elimination of water gives a α, β-unsaturated ketone product.

Not needed here.

126

Now try question 4.21 and 4.22

5 Aromatic chemistry

Bonding in aromatic compounds

The molecular orbital description of bonding in benzene derivatives and the importance of the delocalisation of electrons. The concept of resonance stabilisation and the $(4n+2)$ rule.

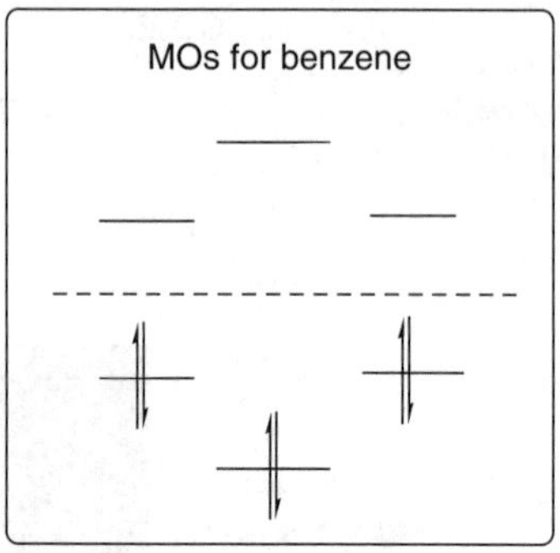

MOs for benzene

The stability and reactivity of benzene derivatives are dominated by their aromaticity and by the fact that the electronic effect of substituents can be transmitted around the ring by resonance:

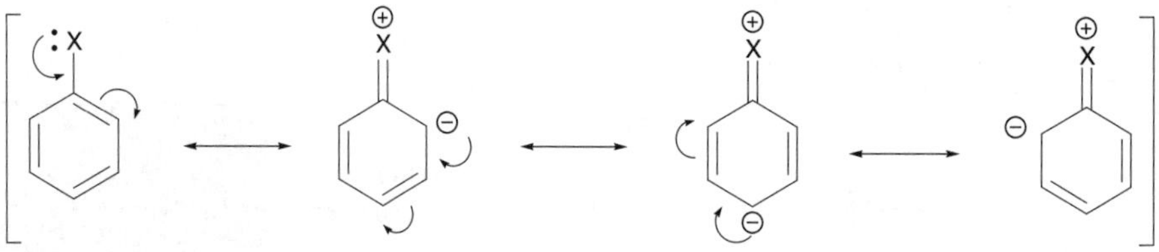

Electrophilic aromatic substitution (SEAr)

(a) *Generalised mechanism*—σ-complexes, π-complexes, the role of Lewis acids, and aromatisation as a driving force for substitution reactions.

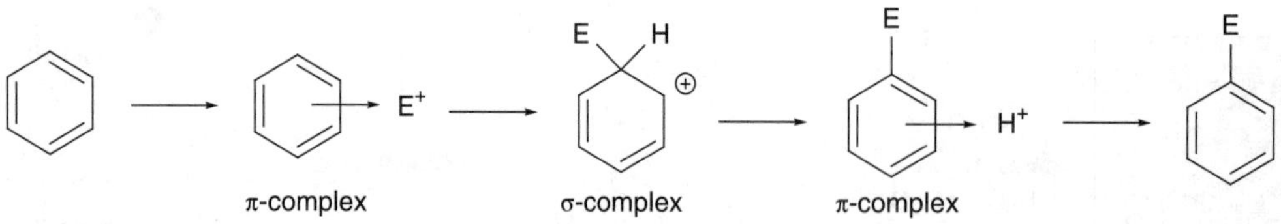

(b) *Substitution of monofunctionalised benzene derivatives*
Orientation and reactivity rules governed by the nature of ring substituents (inductive and mesomeric effects).
The importance of kinetic and thermodynamic control.

(c) *Specific reaction and reagent examples*:
  (i)   Deuteriation: D_2SO_4/D_2O
  (ii) Nitration: HNO_3/H_2SO_4
  (iii) Halogenation: $Br_2/FeBr_3$
  (iv) Sulfonation: SO_3/H_2SO_4

How to Solve Organic Reaction Mechanisms: A Stepwise Approach, First Edition. Mark G. Moloney.
© 2015 John Wiley & Sons, Ltd. Published 2015 by John Wiley & Sons, Ltd. Companion website: www.wiley.com/go/moloney/mechanisms

        (v)   Diazo coupling: PhN_2^+
        (vi)   Friedel–Crafts alkylation and acylation: RCl, $AlCl_3$ and RCOCl, $AlCl_3$
        (vii) Vilsmeier Formylation: $POCl_3$, Me_2NCHO
(d)   *Special case*: the Reimer–Tiemann reaction.

Nucleophilic aromatic substitution (SNAr)

There are three possible mechanisms depending on the reaction conditions and the nature of the Substrate and are as follows:

(a)   *SNAr:* need activation with electron withdrawing groups (Z), usually more than one.

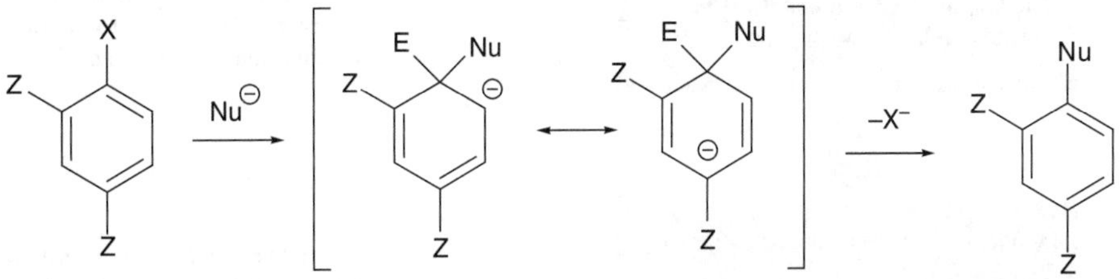

X = leaving group; Z = electron withdrawing group(s)

(b)   *SN1:* diazonium salts react with nucleophiles—the Sandmeyer reaction.

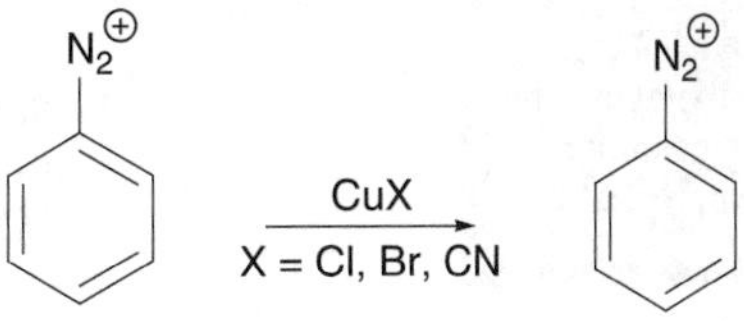

CuX
X = Cl, Br, CN

(c)   *Benzyne mechanism:* aryne intermediates formed by the elimination of aryl halides.

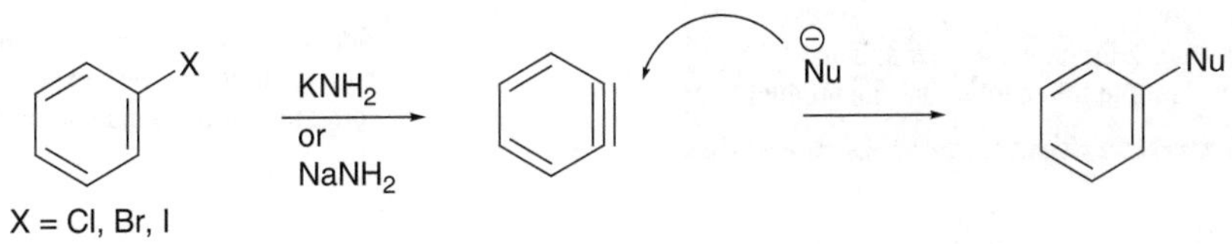

KNH_2
or
$NaNH_2$

X = Cl, Br, I

Oxidation

Alkyl benzenes can be oxidised to benzoic acid, regardless of the chain length, using potassium permanganate.

R

CO_2H

5.1

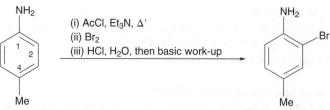

(i) AcCl, Et₃N, Δ'
(ii) Br₂
(iii) HCl, H₂O, then basic work-up

Label electrophilic/nucleophilic and acidic/basic sites of all reactants, and number identical atoms in the starting material and product.

Amines are good nucleophiles (nitrogen possesses a lone pair) and carboxylic acid chlorides are reactive electrophiles by virtue of the polarisation of the carbonyl group due to the electronegativity of chlorine and that Cl⁻ is a good leaving group.

Identify the most reactive sites, if more than one exists.

The aniline nitrogen is the most nucleophilic site, although the lone pair is linked by resonance to the aromatic ring, placing a δ-charge on the *o*- and *p*-positions (labelled C-2 and C-4 here). The acid chloride is the only reactive electrophile.

Recall the characteristic reactions of the most reactive functional groups; and by considering the reaction conditions, decide which is the most appropriate.

Reaction of an acid chloride with an amine nucleophile gives an amide product with expulsion of chloride.

Work through the mechanism leading to the intermediate product.

Addition–elimination of the aniline at the acetyl chloride proceeds with expulsion of chloride and formation of the amide product.

Repeat the above four steps.

* Bromine is a good electrophile, and acetanilide a good nucleophile, due to the presence of the donating amide nitrogen. *Electrophilic aromatic substitution* (S_EAr) by the bromine gives the aryl bromide product, in which substitution of Br⁺ occurs at the *o*-position.
* Acid-catalysed amide *hydrolysis* regenerates the amine, but under the conditions of the hydrolysis reaction this is protonated, to give the anilinium salt as the product.

Recognise that this is not the final product, but is closely related to it.

A basic work-up allows isolation of the free amine.

Write down the structure of the final product.

Summary: This is an example of electrophilic aromatic substitution:

Now try questions 5.8 and 5.9

5.2

 (i) p-TsCl, Na₂CO₃,
then acidic work-up

(ii) PCl₅
(iii) AlCl₃, C₆H₆

 Label electrophilic/nucleophilic and acidic/basic sites of all reactants, and number identical atoms in the starting material and product.

Carboxylic acids and amines are good nucleophiles (both oxygen and nitrogen possess lone pairs). *p*-TsCl is an acid chloride derived from a sulfonic acid, and is electrophilic at the sulfur atom. Sodium carbonate is a mild base.

 Identify the most reactive sites, if more than one exists.

The amine is the most nucleophilic site, since nitrogen is less electronegative than oxygen, even though in this case the carboxylic acid is deprotonated under the basic conditions. *p*-TsCl is the only electrophile present. Sodium carbonate is the only base present.

 Recall the characteristic reactions of the most reactive functional groups; and by considering the reaction conditions, decide which is the most appropriate.

Reaction of a sulfonyl acid chloride with an amine nucleophile gives an amide product, in this case a sulfonamide, with expulsion of chloride.

 Work through the mechanism leading to the intermediate product.

The aromatic amine undergoes a *nucleophilic addition–elimination* reaction at the sulfonyl group, with loss of chloride; this generates a sulfonamide. The basic conditions of this reaction gives a carboxylate product, which is reprotonated on acidic work-up.

 Repeat the above four steps.

* Phosphorus pentachloride is an electrophile, with excellent chloride *leaving groups*. The carboxylic acid reacts with PCl₅ to give an acid chloride, *via* an activated phosphoryl intermediate.
* The acid chloride reacts with the powerful Lewis acid, AlCl₃, to give an *acylium* cation (resonance stabilised), which is very electrophilic and reacts with the good nucleophile, benzene, in an *electrophilic substitution* reaction.

 Recognise that this is not the final product, but is closely related to it.

Loss of a proton regenerates the aromatic system.

Write down the structure of the final product.

Summary: This is an example of the Friedel-Crafts acylation reaction:

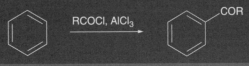

$$\text{RCOCl, AlCl}_3$$

Now try questions 5.10 and 5.16

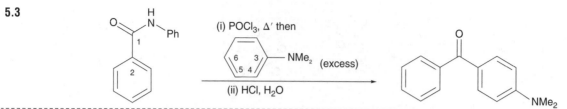

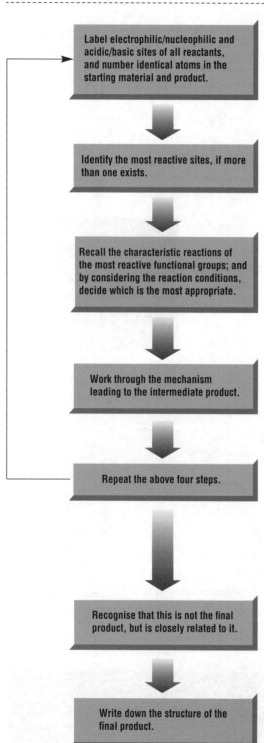

Label electrophilic/nucleophilic and acidic/basic sites of all reactants, and number identical atoms in the starting material and product.

Amines are good nucleophiles, since nitrogen possesses a lone pair. Amides are nucleophilic at oxygen and nitrogen. $POCl_3$ is a potent electrophile, since there are three electron-withdrawing chlorines and each of these is also a good leaving group.

Identify the most reactive sites, if more than one exists.

Amides are most nucleophilic on oxygen (the product has more resonance structures than that from N–alkylation). $POCl_3$ is the only electrophile. Anilines are most nucleophilic at the o- and p-positions, but the latter is more sterically accessible.

Recall the characteristic reactions of the most reactive functional groups; and by considering the reaction conditions, decide which is the most appropriate.

Amides react readily with $POCl_3$ to generate a chloromethyliminium salt; this itself is a potent electrophile, and reacts with aromatic compounds by an electrophilic aromatic substitution (S_EAr) reaction.

Work through the mechanism leading to the intermediate product.

Initial O-phosphorylation of the amide by $POCl_3$ (to form a strong P–O bond) is followed by addition-elimination of the released Cl^- to give a chloromethyliminium ion.

Repeat the above four steps.

* Electrophilic aromatic substitution at the p-position of the aniline with the chloromethyliminium electrophile gives an imine product.
* Acid-catalysed imine hydrolysis, by an addition-elimination mechanism, gives the ketone product.

Recognise that this is not the final product, but is closely related to it.

Final deprotonation of the oxonium cation gives the ketone product.

Write down the structure of the final product.

Summary: This is an example of the Vilsmeier–Haack reaction.

Now try questions 5.11 and 5.17

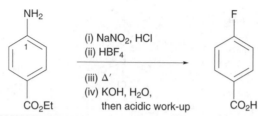

Label electrophilic/nucleophilic and acidic/basic sites of all reactants, and number identical atoms in the starting material and product.

Amines are nucleophiles (oxygen possesses lone pairs). Sodium nitrite under acidic conditions generates HONO. HCl is a strong acid.

Identify the most reactive sites, if more than one exists.

The amine is the most nucleophilic site, although it is deactivated somewhat by the ester substituent at the *p*-position by a resonance delocalisation of the nitrogen lone pair.

Recall the characteristic reactions of the most reactive functional groups; and by considering the reaction conditions, decide which is the most appropriate.

The combination NaNO₂/HCl generates HONO (nitrous acid), which is in *equilibrium* with N_2O_3, a source of $O=N^+$. This reagent *diazotises* amine functional groups, converting NH_2 to N_2^+.

Work through the mechanism leading to the intermediate product.

Nucleophilic addition of the amine to N_2O_3, is followed by *tautomerisation*, which permits *elimination* of H_2O to give the diazonium salt product; this is converted to the BF_4^- salt by treatment with fluoroboric acid.

Repeat the above four steps.

* The diazo group loses nitrogen gas on heating to give an aryl cation; this is intercepted by the available nucleophile, fluoride ion.

* Hydroxide is a good base and nucleophile, and causes alkaline hydrolysis (*addition–elimination* mechanism) of the ester group.

Recognise that this is not the final product, but is closely related to it.

Acidic work-up gives the carboxylic acid product.

Write down the structure of the final product.

Summary: This is an example of the Schiemann reaction, involving the generation of a diazonium cation, followed by attack with F⁻:

Now try questions 5.12 and 5.18

5.5

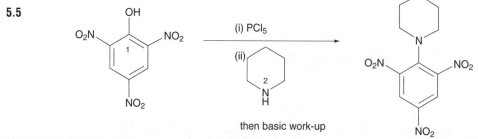

then basic work-up

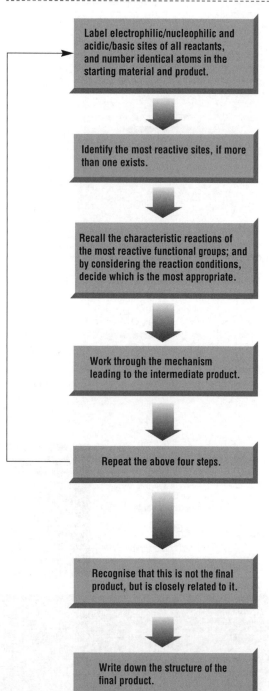

Phenols are nucleophilic at oxygen. PCl$_5$ is an electrophilic chlorinating agent (there are five electron-withdrawing chlorines and each is a good leaving group).

The oxygen of the phenol is the most nucleophilic site, and PCl$_5$ is the only electrophilic agent.

Alcohols react with phosphorus halides to give the corresponding alkyl chloride; aromatic rings activated by electron withdrawing groups are susceptible to aromatic nucleophilic substitution (S$_N$Ar).

Initial attack by the hydroxyl group on phosphorus generates a good leaving group at C-1; a sequence of *addition–elimination* steps then gives the chloride product, giving overall substitution of chloride by hydroxide. This process is facilitated by the presence of electron withdrawing groups which stabilise the intermediated carbanion, and is an example of *aromatic nucleophilic substitution*.

Amines are nucleophilic at nitrogen. The aromatic chloride is highly electron deficient and carries a good *leaving group*; attack by the nucleophilic piperidine then generates the corresponding product by another *addition- elimination* reaction, which is again an example of *aromatic nucleophilic substitution* (S$_N$Ar).

Since the ammonium product is generated under acidic conditions, it is obtained in protonated form. Deprotonation on aqueous work-up gives the final product.

Summary: This is an example of nucleophilic aromatic substitution (S_NAr):

Now try questions 5.13 and 5.19

5.6

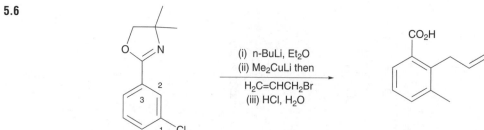

(i) n-BuLi, Et₂O
(ii) Me₂CuLi then
    H₂C=CHCH₂Br
(iii) HCl, H₂O

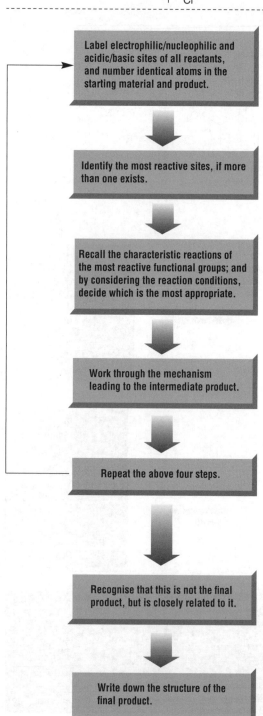

Label electrophilic/nucleophilic and acidic/basic sites of all reactants, and number identical atoms in the starting material and product.	*n*-BuLi is a strong base (pK$_a$ of butane = 50). The aromatic protons α- to the chloro substituent at C-2 are weakly acidic. Dimethyllithium cuprate is a source of methyl anion, which is highly nucleophilic.
Identify the most reactive sites, if more than one exists.	The proton at C-2 is the most acidic, since the corresponding lithiated species is stabilised by chelation to the adjacent nitrogen atom.
Recall the characteristic reactions of the most reactive functional groups; and by considering the reaction conditions, decide which is the most appropriate.	Aromatic halides readily react with strong bases to give a carbanion, especially when it is stabilised by adjacent substituents. This carbanion can spontaneously eliminate, to give a *benzyne* intermediate.
Work through the mechanism leading to the intermediate product.	Removal of the proton at C-2, giving a carbanion species in which the lithium is stabilised by chelation to the adjacent nitrogen and chlorine atoms, is followed by β- elimination of chloride. This gives the benzyne intermediate directly.
Repeat the above four steps.	* Benzyne is highly electrophilic, and is readily attacked by nucleophiles, in this case dimethylcuprate. The resulting anion is trapped by allyl bromide (as an electrophile) in a *nucleophilic substitution* reaction. * The oxazoline intermediate is hydrolysed with dilute acid in a series of addition–elimination steps to give the corresponding carboxylic acid product.
Recognise that this is not the final product, but is closely related to it.	Not needed here.
Write down the structure of the final product.	

Summary: This is an example of aromatic substitution involving a benzyne intermediate

Now try questions 5.14 and 5.20

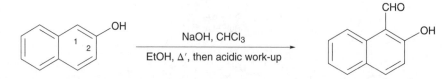

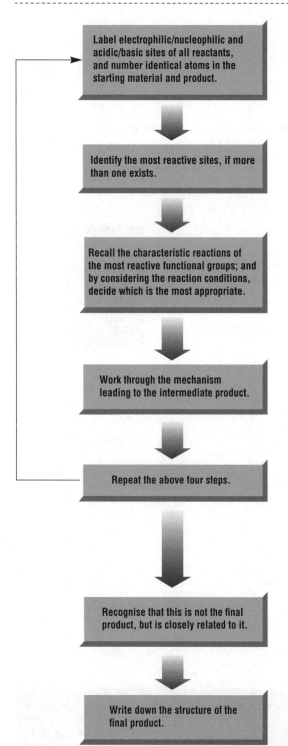

2-Naphthol is a highly nucleophilic aromatic compound, activated by resonance release from the phenolic oxygen. NaOH is a strong base, and CHCl₃ possesses an acidic hydrogen, since the carbanion is stablised by electron withdrawl from the adjacent chlorine atoms.

C-1 of 2-naphthol is especially activated (*o*- to the hydroxy group) to electrophilic attack. NaOH is the only base, and both naphthol and CHCl₃ possess acidic hydrogens.

NaOH deprotonates CHCl₃, and will give *dichlorocarbene* by an α-*elimination* reaction. This intermediate is very electrophilic, and will rapidly react with suitable nucleophiles. Under these conditions, naphthol is also deprotonated to the alkoxide, and this makes it very susceptible to *electrophilic aromatic substitution*.

Electrophilic aromatic substitution at C-1 of 2-naphthol with dichlorocarbene gives the dichloromethyl product. Elimination of chloride generates an enone which is intercepted by hydroxide; this regenerates the aromatic ring. Collapse of the intermediate so formed generates the aldehyde product.

Not needed here.

Acidic work-up reprotonates the phenoxide anion, giving the product.

Summary: This is an example of the Reimer–Tiemann reaction:

NaOH, CHCl₃

Now try questions 5.15 and 5.21

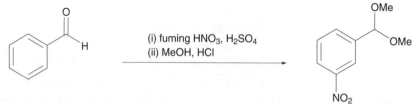

The carbonyl group is polarised by the *electronegativity* difference between carbon and oxygen, making the carbon electophilic. The aromatic ring is nucleophilic, but deactivated by the presence of the aldehyde group. Nitric acid/sulfuric acid is a strong acid mixture, which generates NO_2^+.

Label electrophilic/nucleophilic and acidic/basic sites of all reactants, and number identical atoms in the starting material and product.

The aldehyde is the only electrophilic site, and the aromatic ring the only nucleophile. The nitric/sulfuric acid mixture generates NO_2^+, which is a potent electrophile.

Identify the most reactive sites, if more than one exists.

Aromatic rings are very susceptible to *electrophilic aromatic substitution* (S_EAr), although in this case, the aldehyde group is deactivating and *m*-directing.

Recall the characteristic reactions of the most reactive functional groups; and by considering the reaction conditions, decide which is the most appropriate.

Electrophilic aromatic substitution at the *m*-position of benzaldehyde with the NO_2^+ electrophile gives the product, *m*-nitrobenzaldehyde.

Work through the mechanism leading to the intermediate product.

The aldehyde function is very electrophilic; protonation of the carbonyl groups leads to a sequence of addition–elimination steps involving addition of methanol which gives an acetal product. Final deprotonation of the *oxonium* cation gives the ketone product.

Repeat the above four steps.

Recognise that this is not the final product, but is closely related to it.

Not needed here

Write down the structure of the final product.

Summary: This is an example of electrophilic aromatic substitution:

Now try question 5.9

5.9

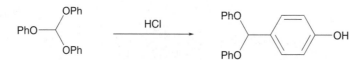

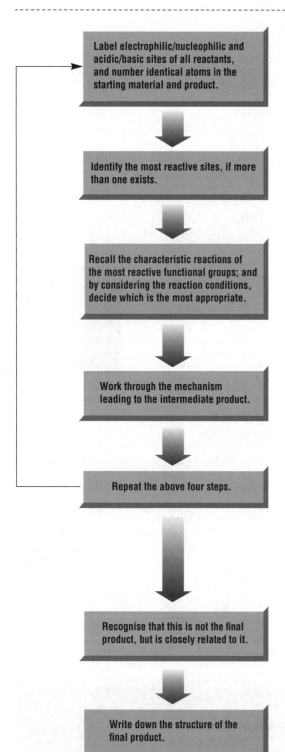

Label electrophilic/nucleophilic and acidic/basic sites of all reactants, and number identical atoms in the starting material and product.

Orthoesters are basic, since the oxygen atoms have lone pairs able to accept a proton. HCl is a strong acid and fully ionised ($pK_a = -7$).

Identify the most reactive sites, if more than one exists.

This orthoester is symmetrical, and so any one of the oxygen atoms may be protonated by the strong acid, HCl.

Recall the characteristic reactions of the most reactive functional groups; and by considering the reaction conditions, decide which is the most appropriate.

Orthoesters, like acetals, readily collapse under acidic conditions.

Work through the mechanism leading to the intermediate product.

Orthoesters are like acetals, in that protonation of one of the oxygen atoms leads to the formation of a leaving group, whose departure may be assisted by the adjacent lone pair. In this case, this generates an *oxonium* cation and phenol.

Repeat the above four steps.

Aromatic rings are very susceptible to *electrophilic aromatic substitution* (S_EAr), and in this case, the hydroxy group is strongly activating and *o, p*-directing. Electrophilic aromatic substitution at the less hindered *p*-position of phenol with the generated electrophile gives the product.

Recognise that this is not the final product, but is closely related to it.

Not needed here.

Write down the structure of the final product.

Summary: This is an example of electrophilic aromatic substitution:

147

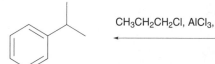 Me₃CCl, AlCl₃,

CH₃CH₂CH₂Cl, AlCl₃,

Label electrophilic/nucleophilic and acidic/basic sites of all reactants, and number identical atoms in the starting material and product.

Benzene is a good nucleophile, by virtue of its π-electron density. Alkyl chlorides are activated by the potent Lewis acid, aluminium chloride, to departure of the chloride leaving group, to give an intermediate carbocation.

Identify the most reactive sites, if more than one exists.

Aluminium chloride and the alkyl halides are both electrophiles, and the aromatic ring the only nucleophile.

Recall the characteristic reactions of the most reactive functional groups; and by considering the reaction conditions, decide which is the most appropriate.

Aromatic rings are very susceptible to *electrophilic aromatic substitution* (S_EAr), which in this case is simply benzene.

Work through the mechanism leading to the intermediate product.

Aluminium chloride accepts a chloride leaving group from both of the alkyl chloride starting materials, to generate $AlCl_4^-$ and the corresponding carbocation. The *t*-butyl carbocation being tertiary is stabilised and able to react further with benzene. In the case of the *n*-propyl carbocation, the first formed primary carbocation readily rearranges by 1,2-hydrogen shift to give a more stabilised secondary *iso*-propyl carbocation and it is this which reacts further, again with benzene. Electrophilic aromatic substitution of benzene then gives each of the two products.

Repeat the above four steps.

Recognise that this is not the final product, but is closely related to it.

Not needed here.

Write down the structure of the final product.

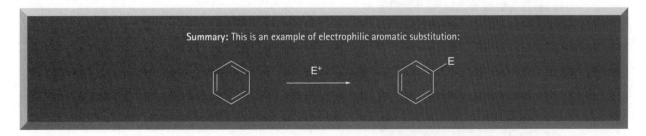

Summary: This is an example of electrophilic aromatic substitution:

Now try question 5.16

$$\xrightarrow{\text{POCl}_3, \Delta'}$$

Label electrophilic/nucleophilic and acidic/basic sites of all reactants, and number identical atoms in the starting material and product.

Amides are nucleophilic at oxygen and nitrogen. An aromatic ring possessing electron-releasing groups, such as the methoxy groups here, is also nucleophilic. POCl₃ is a potent electrophile, since there are three electron-withdrawing chlorines and each of these is also a good leaving group.

Identify the most reactive sites, if more than one exists.

Amides are most nucleophilic on oxygen (the product has more resonance structures than that from *N*-alkylation). POCl₃ is the only electrophile. The methoxy groups of the aromatic ring are activating towards nucleophilic substitution at the *o*- and *p*-positions, but the latter is not accessible in this case.

Recall the characteristic reactions of the most reactive functional groups; and by considering the reaction conditions, decide which is the most appropriate.

Amides react readily with POCl₃ to generate a chloromethyliminium salt; this itself is a potent electrophile, and reacts with aromatic compounds by an electrophilic aromatic substitution (*S*ₑ*Ar*) reaction.

Work through the mechanism leading to the intermediate product.

Initial *O*-phosphorylation of the amide by POCl₃ (to form a strong P–O bond) is followed by *addition–elimination* of the released Cl⁻ to give a chloromethyliminium ion. This is not isolated but reacts immediately onwards.

Repeat the above four steps.

Electrophilic aromatic substitution at the *o*-position of one of the methoxy groups with the chloromethyliminium electrophile gives an imine product, in which the six membered ring has been formed. Final deprotonation of the *iminium* cation gives the imine product.

Recognise that this is not the final product, but is closely related to it.

Not needed here.

Write down the structure of the final product.

Now try question 5.17

(i) PCl$_5$

(ii) NH$_3$

(iii) ^{t}BuON=O then NaBF$_4$

Label electrophilic/nucleophilic and acidic/basic sites of all reactants, and number identical atoms in the starting material and product.

Amines are good nucleophiles (nitrogen possesses a lone pair); this amine is linked by resonance to the ketone, and is called a pyridone. PCl$_5$ is an electrophilic chlorinating agent (there are five electron-withdrawing chlorines and each is a good leaving group).

Identify the most reactive sites, if more than one exists.

The amine is the most nucleophilic site, but as a result of resonance delocalisation of the nitrogen lone pair, the oxygen is also nucleophilic. Phosphorus pentachloride is an electrophile, with excellent chloride *leaving groups*.

Recall the characteristic reactions of the most reactive functional groups; and by considering the reaction conditions, decide which is the most appropriate.

Alcohols react with phosphorus halides to give the corresponding alkyl chloride; aromatic rings activated by electron withdrawing groups are susceptible to aromatic nucleophilic substitution (S$_N$Ar).

Work through the mechanism leading to the intermediate product.

The pyridone reacts with PCl$_5$ to give an activated phosphoryl intermediate, which is susceptible to attack by chloride; addition–elimination gives the chloro product.

Repeat the above four steps.

* The chloride is a good leaving group and susceptible to S$_N$Ar, by addition–elimination using ammonia as the nucleophile.

* *t*-BuONO is a source of O=N$^+$, by reacting with amines in a nucleophilic addition–elimination sequence. This reagent *diazotises* the amine functional groups, converting NH$_2$ to N$_2^+$, with loss of *t*-BuO$^-$. This is followed by *automerisation*, which permits *elimination* of H$_2$O to give the diazonium salt product.

* The diazo group loses nitrogen gas on heating to give an aryl cation which is strongly electrophilic; this is intercepted by the available nucleophile, fluoride ion.

Recognise that this is not the final product, but is closely related to it.

Write down the structure of the final product.

Not needed here.

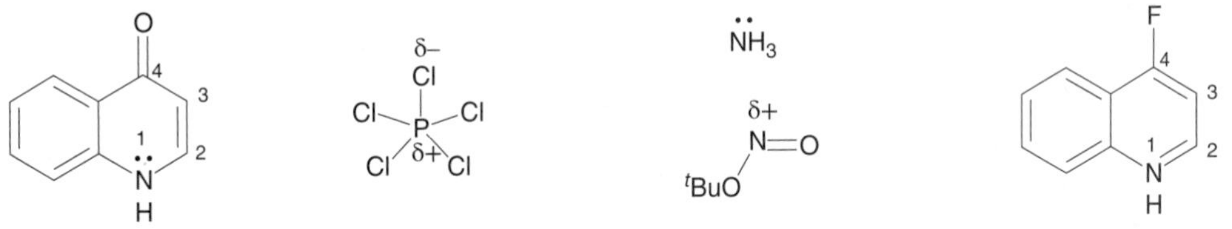

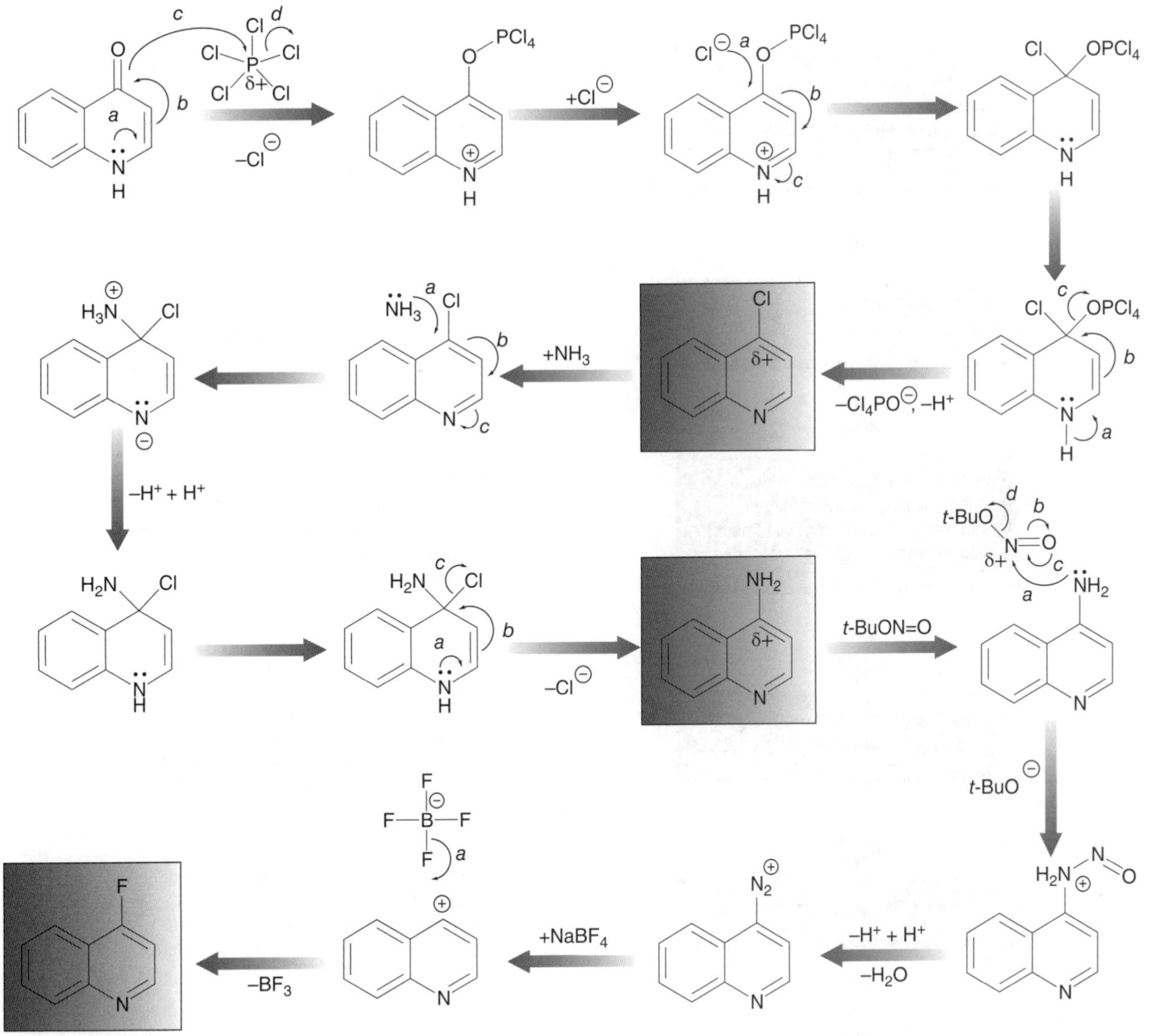

O
4
3
1
2
N
H

δ−
Cl
Cl   Cl
P
δ+
Cl   Cl

$\cdot\cdot$
NH_3
δ+
^tBuON=O

F
4
3
1
2
N
H

Summary: This is an example of nucleoophilic aromatic substitution:

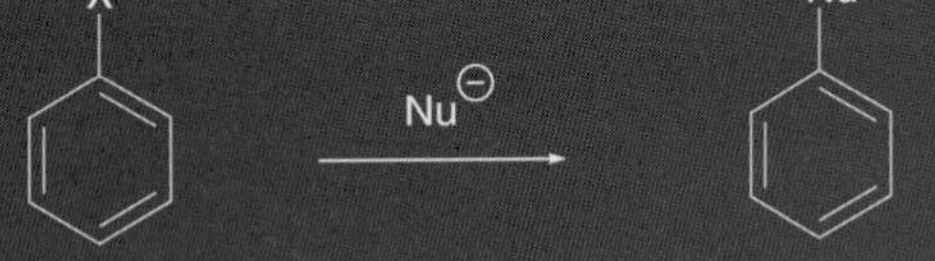

X
Nu$^{\ominus}$
Nu

Now try question 5.18

5.13

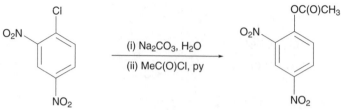

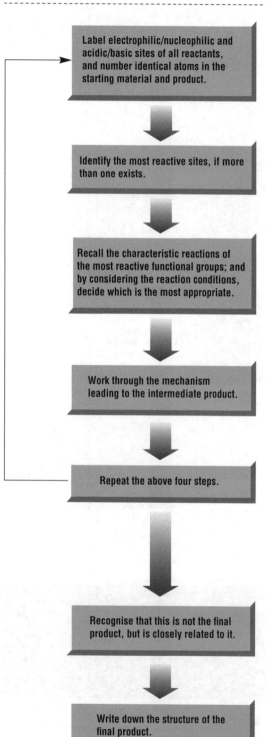

Aryl chlorides substituted with strongly electron withdrawing groups are electrophiles, since chlorine is electronegative and chloride is a good leaving group. Sodium carbonate is a weak base but also a good nucleophile.

The aryl chloride is the only electrophilic site and carbonate is the only nucleophile.

The aromatic chloride is highly electron deficient and carries a good *leaving group*; it will readily participate in *aromatic nucleophilic substitution* (S$_N$Ar).

Attack by the nucleophilic carbonate generates the corresponding product by *addition-elimination* reaction, leading to overall aromatic nucleophilic substitution. This product will spontaneously lose carbon dioxide, and the resulting phenolate is protonated on work-up.

Reaction of the phenol with acetyl chloride proceeds by an addition-elimination mechanism at the carbonyl group, which leads to esterification by expulsion of chloride.

Not needed here.

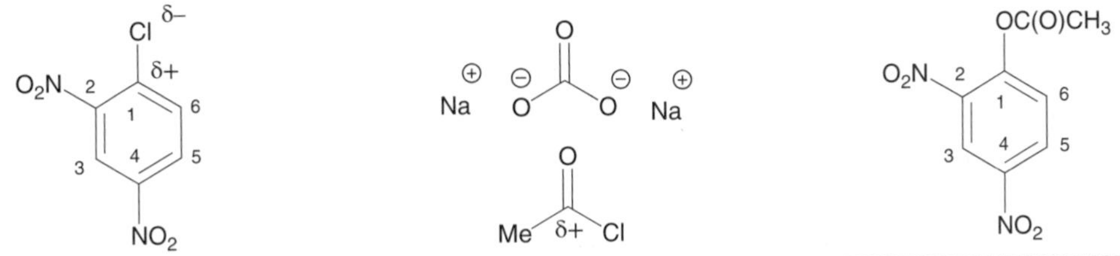

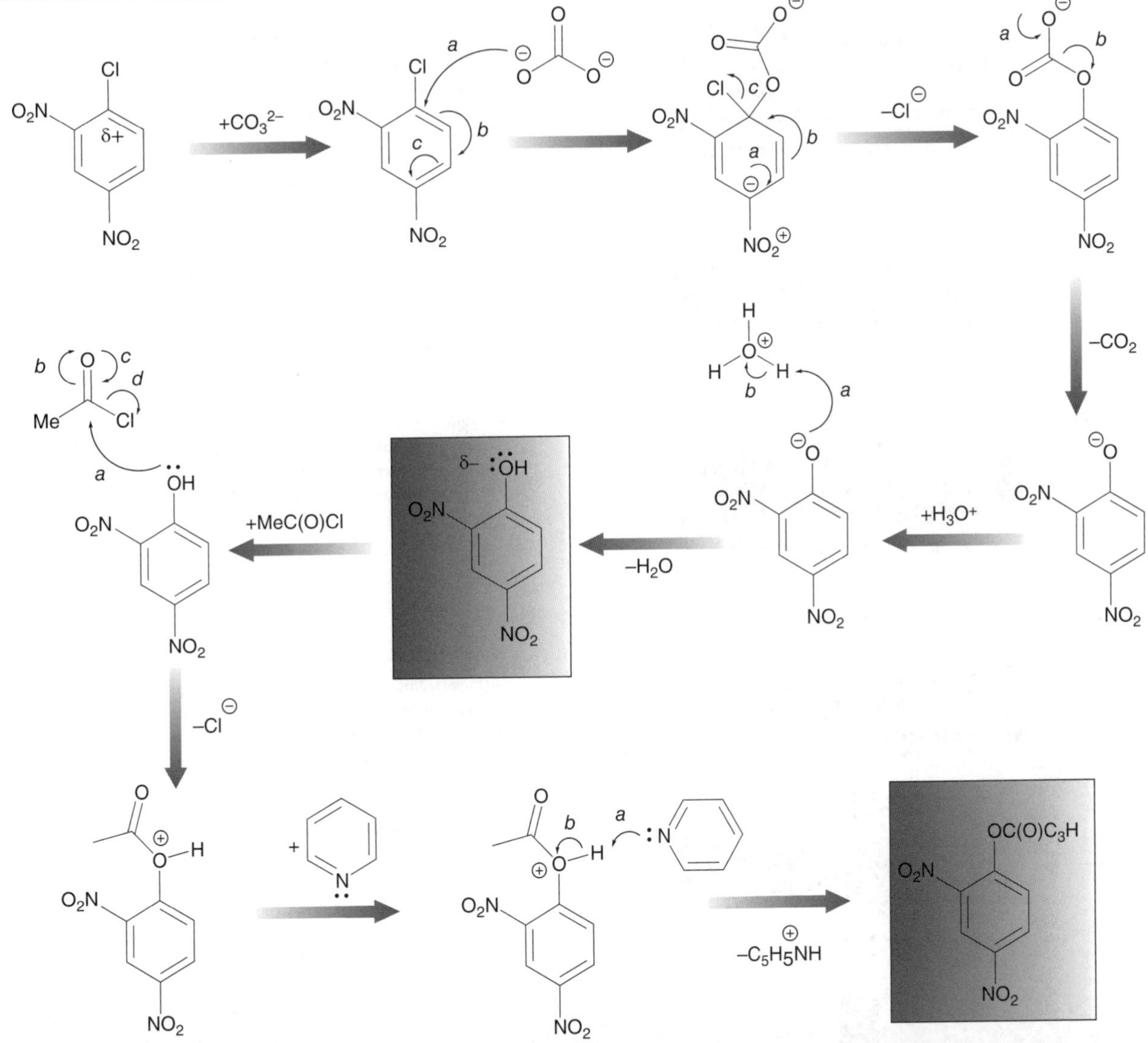

Summary: This is an example of nucleophilic aromatic substitution (S$_N$Ar):

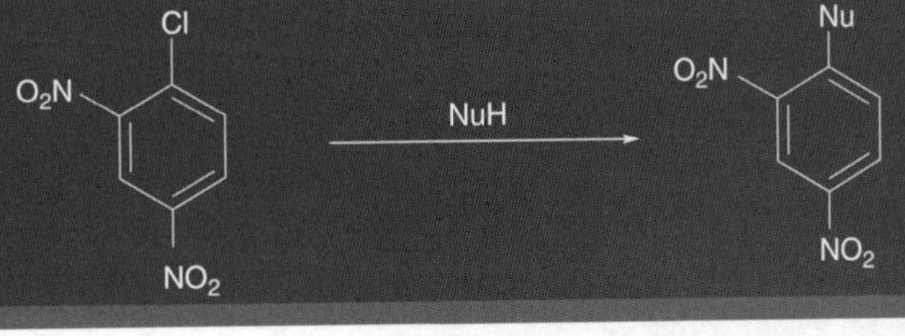

Now try question 5.19

155

5.14

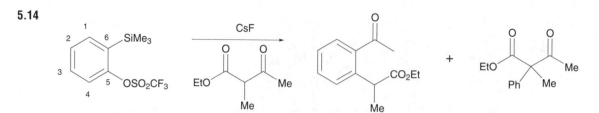

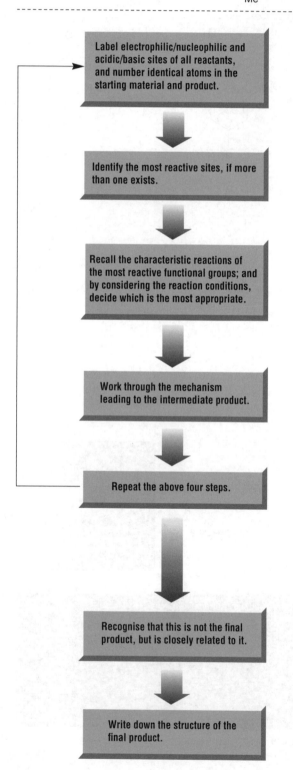

The carbonyl group is polarised by the *electronegativity* difference between carbon and oxygen, making the carbon electrophilic. The α-protons of a carbonyl group are acidified by *inductive* withdrawl, and this allows formation of the resulting enol which is a good nucleophile. Trifluoromethanesulfonate is an excellent leaving group. Fluoride is a good nucleophile and a weak base.

The silyl group is the most electrophilic site and silicon forms very strong bonds with fluoride. The β-dicarbonyl is the most acidic reagent (pK_a 13) and fluoride is a weak base; it exists in equilibrium with its enolic form.

Aromatic compounds with good leaving groups may eliminate to give a *benzyne* intermediate, which readily reacts with nucleophiles.

Attack by fluoride at silicon initiates collapse to benzyne by loss of triflate. This is intercepted by the enol which generates a carbanion. This carbanion may be protonated immediately, giving one of the observed products, or may undergo rearrangement with acyl group transfer, leading to formation of an enolate, which after protonation gives the other product.

Not needed here.

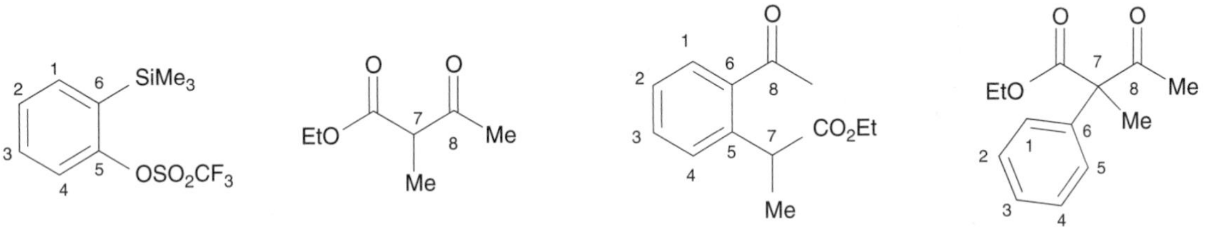

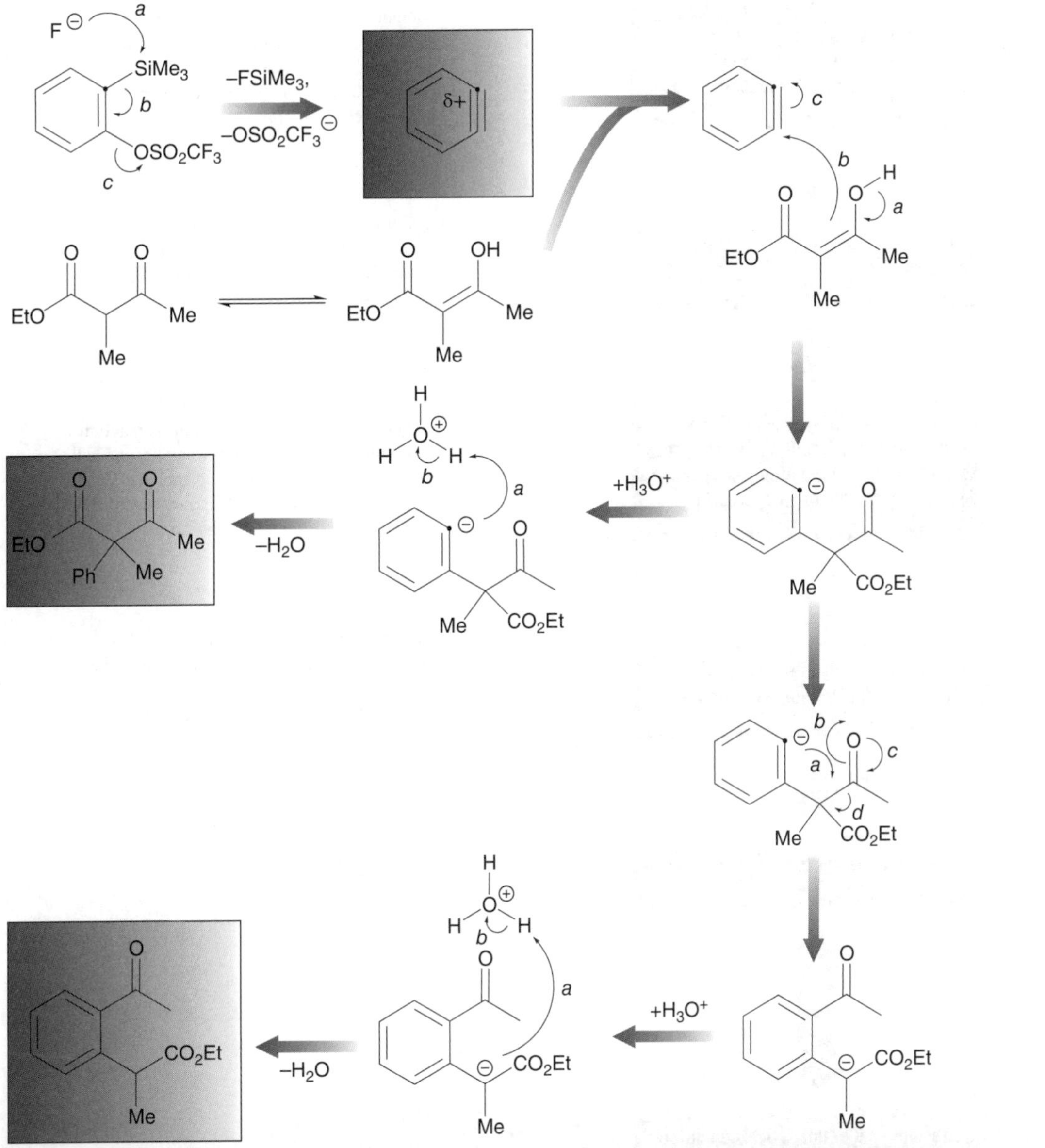

Summary: This is an example of aromatic substitution involving a benzyne intermediat:

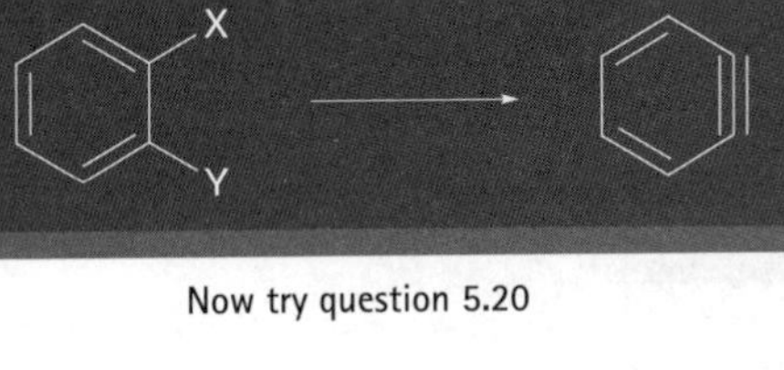

Now try question 5.20

157

5.15

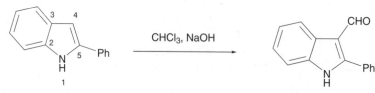

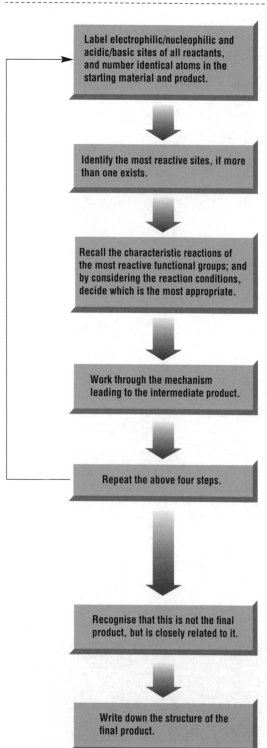

Label electrophilic/nucleophilic and acidic/basic sites of all reactants, and number identical atoms in the starting material and product.	Indoles are highly nucleophilic aromatic compounds, activated by the nitrogen. NaOH is a strong base, and CHCl$_3$ possesses an acidic hydrogen, since the carbanion is stablised by electron withdrawal from the adjacent chlorine atoms.
Identify the most reactive sites, if more than one exists.	C-4 of the indole is especially activated (β- to the amine group) to electrophilic attack. NaOH is the only base, and only CHCl$_3$ possesses acidic hydrogens.
Recall the characteristic reactions of the most reactive functional groups; and by considering the reaction conditions, decide which is the most appropriate.	NaOH deprotonates CHCl$_3$, and will give *dichlorocarbene* by an α-*elimination* reaction. This intermediate is very electrophilic, and will rapidly react with suitable nucleophiles. Under these conditions, the indole is very susceptible to *electrophilic aromatic substitution*.
Work through the mechanism leading to the intermediate product.	*Electrophilic aromatic substitution* at C–4 of the indole with dichlorocarbene gives the dichloromethyl product. Attack by hydroxde and collapse of the intermediate so formed generates the aldehyde product.
Repeat the above four steps.	Not needed here.
Recognise that this is not the final product, but is closely related to it.	Not needed here.
Write down the structure of the final product.	

Summary: This is an example of electrophilic aromatic substitution of pyrrole:

Now try question 5.21

6 Rearrangements

Rearrangements most commonly occur in compounds which possess an electron releasing group (X), a migrating group (R) and a leaving group (L); the ability for correct orbital overlap is critical for the reaction pathway to occur, and there is a clear preference for an *anti*-relationship between migrating and leaving groups. Aryl and alkyl shifts are generally preferred over hydrogen shifts, with aryl more likely than alkyl. The reaction can be represented in its broadest form as follows:

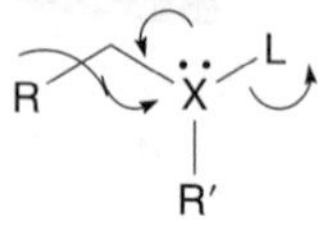

There are several types:

(a) *1,2 Alkyl or aryl shifts to carbon:* often driven by the loss of an appropriate leaving group.
Examples: Wagner–Meerwein rearrangement—a process in which a carbocation rearranges by a 1,2-migration of a hydrogen, alkyl or aryl group.

Pinacol–pinacolone

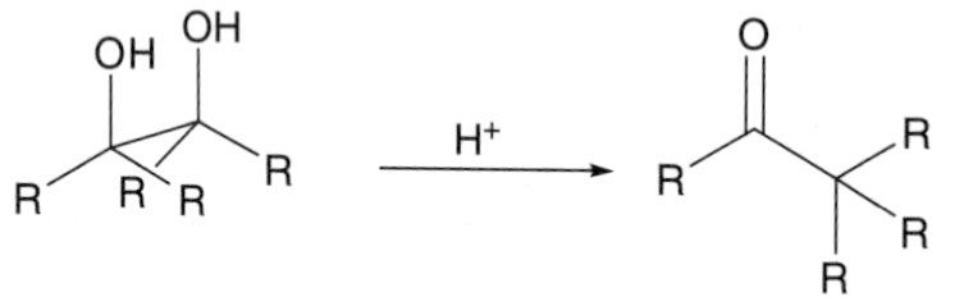

Benzil–benzilic acid

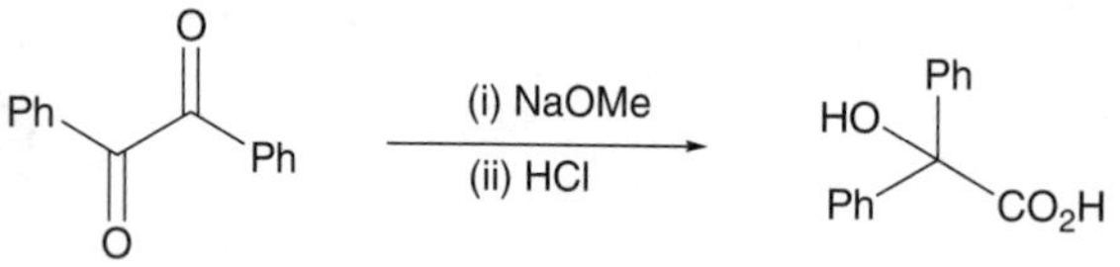

Favorskii

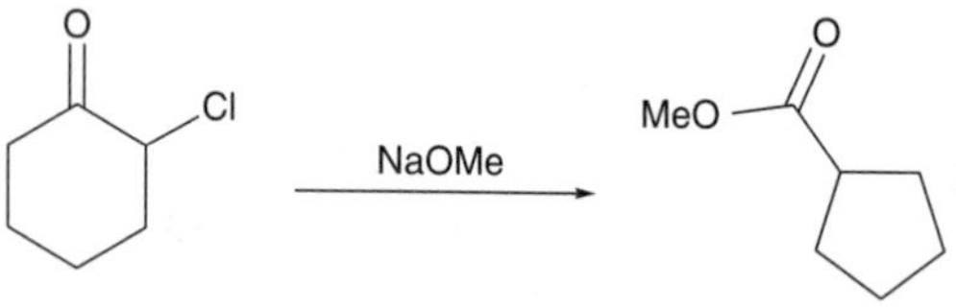

Tiffeneau–Demyanov

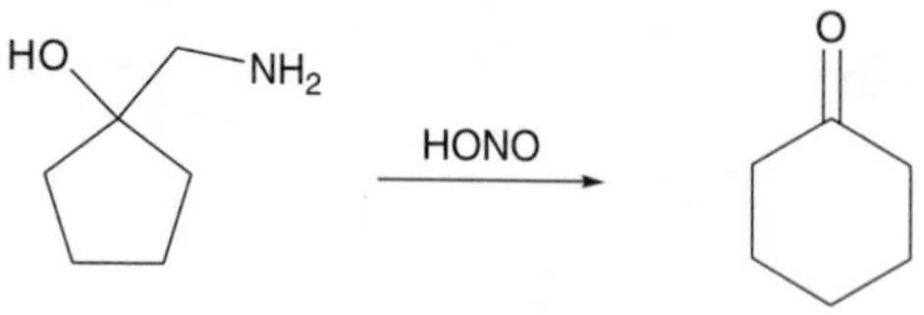

Dienone–phenol

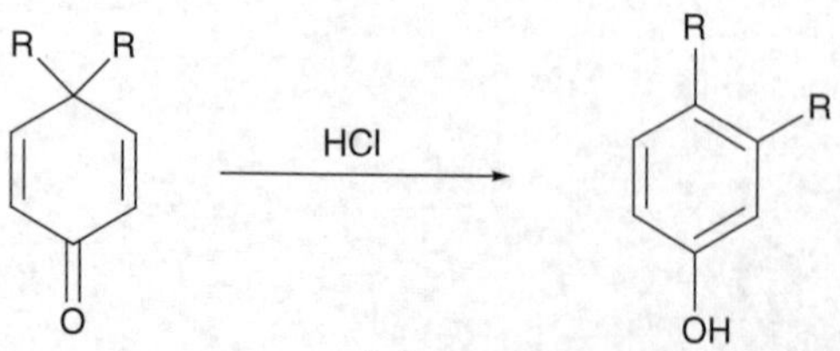

How to Solve Organic Reaction Mechanisms: A Stepwise Approach, First Edition. Mark G. Moloney.
© 2015 John Wiley & Sons, Ltd. Published 2015 by John Wiley & Sons, Ltd. Companion website: www.wiley.com/go/moloney/mechanisms

Wolff (Arndt–Eistert reaction)

$$RCOCl \xrightarrow[\text{(ii) Ag}_2\text{O, R'OH}]{\text{(i) CH}_2\text{N}_2} RCH_2CO_2R'$$

(b) *1,2 Alkyl or aryl shifts to nitrogen or oxygen*
Examples: Beckmann (reaction of oximes with PCl$_5$ or strong acid)

Hofmann (reaction of a primary amide with alkaline bromine solution)

Schmidt (reaction of carboxylic acids with HN$_3$)

Lossen (reaction of an O-acyl hydroxamic acid with formation of an O-leaving group)

Curtius (pyrolysis of acyl azides)

Baeyer–Villiger (reaction of ketones with peracids)

$$RCOR' \xrightarrow{R''CO_3H} RCO_2R'$$

Hydroperoxide (reaction of hydroperoxides under acidic conditions)

(c) *Hydride transfer reactions*
Examples: Cannizzaro (reaction of aromatic aldehydes under alkaline conditions)

Meerwein–Ponndorf–Verley reduction and Oppenauer oxidation (disproportionation of a ketone/isopropyl alcohol or alcohol/acetone mixture respectively in the presence of aluminium isopropoxide).

6.1

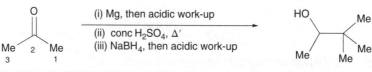

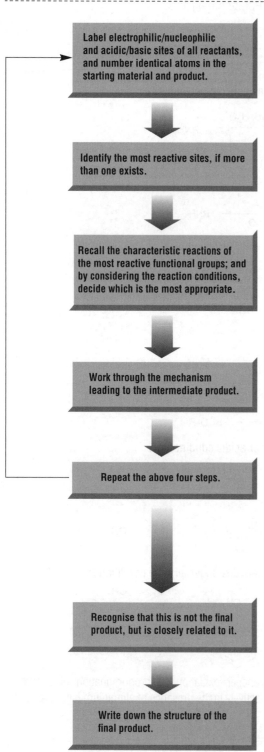

The carbonyl group is polarised by the *electronegativity* difference between carbon and oxygen, making the carbon electrophilic. Magnesium is a reducing metal, and can donate two electrons, in two separate single electron processes.

The ketone is the only reactive electrophilic site, and magnesium the only reducing metal.

Reducing metals (such as Mg) can couple ketones to give 1,2-diols; this is called the Pinacol reaction.

Magnesium donates a single electron to the oxygen of the C=O bond, to form a stable O-Mg bond, and a carbon centred *radical*; this occurs twice to give a magnesium bis(alkoxide). The radicals then *dimerise*, and the bisalkoxide product is protonated on work-up to give a diol.

* Alcohols are basic and nucleophilic. Treatment with strong acid (H_2SO_4) protonates one of the hydroxyl groups, and this induces a *rearrangement*: the lone pair of the other hydroxyl pushes in, forces a 1,2-methyl group migration of the methyl group *anti* to the leaving group, followed by departure of the *leaving group* (H_2O).

* $NaBH_4$ is a source of (nucleophilic) hydride. The carbonyl group is polarised by the *electronegativity* difference between carbon and oxygen, making the carbon electrophilic. *Nucleophilic addition* of hydride generates an alkoxide anion.

Protonation on work–up generates the alcohol product.

Summary: This is an example of the Pinacol–Pinacolone rearrangement:

Now try questions 6.8 and 6.9

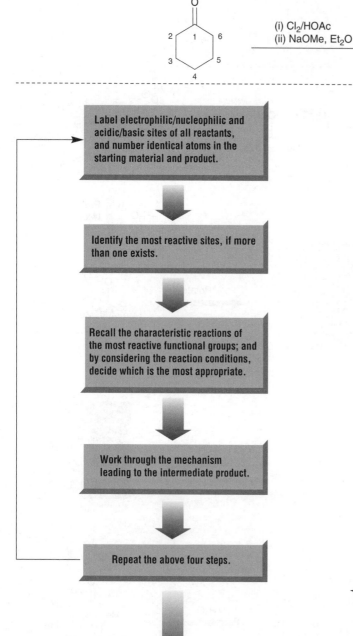

(i) Cl₂/HOAc

Label electrophilic/nucleophilic and acidic/basic sites of all reactants, and number identical atoms in the starting material and product.

The carbonyl group is polarised by the *electronegativity* difference between carbon and oxygen, making the carbon electrophilic. Chlorine is an electrophile, with a weak Cl–Cl bond. HOAc is a weak acid.

Identify the most reactive sites, if more than one exists.

The ketone is the only reactive group, which by tautomerisation is in equilibrium with its enol form in a process catalysed by acid. Chlorine is the only reactive electrophile.

Recall the characteristic reactions of the most reactive functional groups; and by considering the reaction conditions, decide which is the most appropriate.

Ketones are readily halogenated at the α-position by reaction of their corresponding enol (easily generated under acidic conditions by *tautomerisation*).

Work through the mechanism leading to the intermediate product.

The α-position of the enol of cyclohexanone (here C-6) attacks chlorine, to give an α-chloroketone.

Repeat the above four steps.

* Sodium methoxide is a strong base. Deprotonation at C-2 gives an enolate, and this undergoes an intramolecular *nucleophilic substitution* at the carbon bearing the chlorine, to generate a cyclopropanone.

* Cyclopropanones are particularly susceptible to *nucleophilic addition* due to *ring strain*. Methoxide is a good base and nucleophile, and undergoes an *addition–elimination* reaction at the carbonyl with concomitant ring opening.

Recognise that this is not the final product, but is closely related to it.

Protonation of the alkyl anion by methanol gives the product.

Write down the structure of the final product.

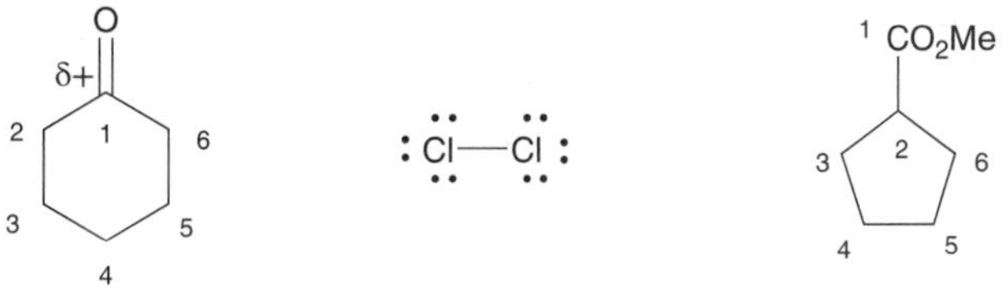

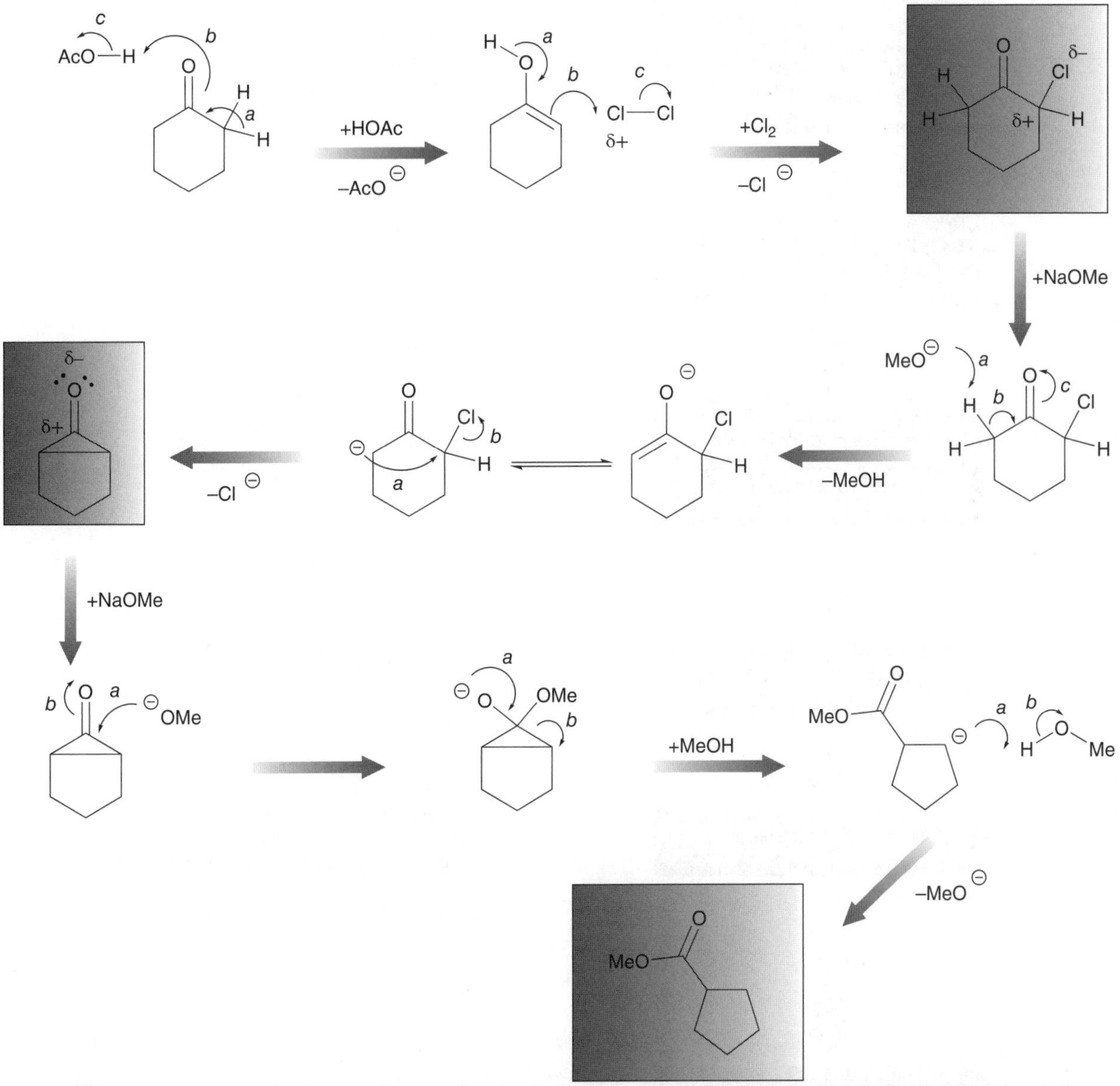

$\delta+$

O

2 1 6

3 5

4

:Cl—Cl:

^{1} CO$_2$Me

3 2 6

4 5

---

c *b*

AcO—H

a

+HOAc

−AcO$^{\ominus}$

H O *a*

b

c

Cl—Cl

$\delta+$

+Cl$_2$

−Cl$^{\ominus}$

O $\delta-$

H Cl

H $\delta+$ H

+NaOMe

MeO$^{\ominus}$ *a*

O *c*

H *b* Cl

H H

$\delta-$ O

$\delta+$

−Cl$^{\ominus}$

O Cl *b*

$^{\ominus}$ *a* H

O$^{\ominus}$ Cl

H

−MeOH

+NaOMe

b O *a*

$^{\ominus}$OMe

$^{\ominus}$ O OMe *a*

b

+MeOH

MeO O

$^{\ominus}$

a *b* H O Me

−MeO$^{\ominus}$

MeO O

Summary: This is an example of the Favorskii rearrangement:

O

Cl

NaOMe

MeO O

Now try questions 6.10 and 6.16

165

6.3

(i) NH$_2$OH
(ii) H$_2$SO$_4$, H$_2$O, Δ'

| Label electrophilic/nucleophilic and acidic/basic sites of all reactants, and number identical atoms in the starting material and product. | The carbonyl group is polarised by the *electronegativity* difference between carbon and oxygen, making the carbon electrophilic. Hydroxylamine is a very good nucleophile (two adjacent heteroatoms, both with lone pairs). |

| Identify the most reactive sites, if more than one exists. | The ketone is the only reactive electrophilic group, and hydroxylamine is the only nucleophile. |

| Recall the characteristic reactions of the most reactive functional groups; and by considering the reaction conditions, decide which is the most appropriate. | Hydroxylamine and ketones react with elimination of water to give *oximes*. |

| Work through the mechanism leading to the intermediate product. | *Addition–elimination* of hydroxylamine to the ketone with expulsion of water, gives the oxime product. |

| Repeat the above four steps. | H$_2$SO$_4$ is a strong acid and fully ionised (pK$_a$ = –9), and oximes are basic on oxygen. Protonation of the *oxime* generates a *leaving group*; the adjacent alkyl group migrates with departure of the leaving group to give a vinylic *carbocation*. This is intercepted by water, to give the product, an amide in its enolic form. |

| Recognise that this is not the final product, but is closely related to it. | *Tautomerisation* to the keto form gives the amide product. |

| Write down the structure of the final product. |

Summary: This is an example of the Beckmann rearrangement:

Now try questions 6.11 and 6.17

6.4

(i) CH_3NO_2, NaOEt
(ii) H_2, Raney Nickel, HOAc
(iii) $NaNO_2$, HOAc, H_2O, 0°C

Label electrophilic/nucleophilic and acidic/basic sites of all reactants, and number identical atoms in the starting material and product.

The carbonyl group is polarised by the *electronegativity* difference between carbon and oxygen, making the carbon electrophilic. Nitromethane is a strong acid (pK$_a$ = 10), easily deprotonated by sodium ethoxide, to give a *nitronate* anion.

Identify the most reactive sites, if more than one exists.

The ketone is the only reactive electrophilic group, and nitromethane which, after deprotonation, gives a nitronate anion, which is a nucleophile.

Recall the characteristic reactions of the most reactive functional groups; and by considering the reaction conditions, decide which is the most appropriate.

The carbon of the ketone is susceptible to *nucleophilic addition*.

Work through the mechanism leading to the intermediate product.

Nucleophilic addition of the nitronate anion to the carbonyl group gives an alkoxide; protonation on work-up gives the corresponding alcohol.

Repeat the above four steps.

* Raney Nickel is a good reducing agent, capable of converting a nitro group to an amine (the mechanism is not well defined, but leads to the addition of hydrogen).
* *Diazotisation* with nitrous acid generates a very good leaving group (nitrogen gas). The oxygen lone pair pushes in, forces a *1,2-alkyl shift* with loss of nitrogen, to give the intermediate product, which is the protonated form of cycloheptanone.

Recognise that this is not the final product, but is closely related to it.

Deprotonation to the keto form gives the product, cycloheptanone.

Write down the structure of the final product.

Summary: This is an example of the Tiffenau–Demjanov Rearrangement:

Now try questions 6.12 and 6.18

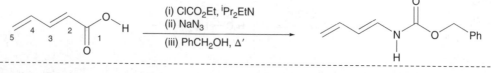

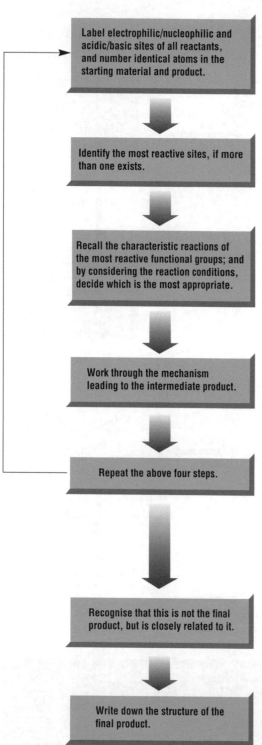

Label electrophilic/nucleophilic and acidic/basic sites of all reactants, and number identical atoms in the starting material and product.	A carboxylic acid is electrophilic at the carbonyl carbon, and nucleophilic at the oxygen atoms. A chloroformate is electrophilic at the carbonyl carbon, and carries two leaving groups, of which chloride is the better. ^{i}Pr$_2$EtN is a good non-nucleophilic base.
Identify the most reactive sites, if more than one exists.	The chloroformate is the more reactive electrophile due to the presence of a highly electron-withdrawing chlorine, and the carboxylic acid may act as a nucleophile. ^{i}Pr$_2$EtN is the only base
Recall the characteristic reactions of the most reactive functional groups; and by considering the reaction conditions, decide which is the most appropriate.	Chloroformates are highly reactive to nucleophilic addition–elimination.
Work through the mechanism leading to the intermediate product.	*Addition–elimination* of the nucleophilic carboxylic acid to the chloroformate gives a mixed anhydride after expulsion of chloride.
Repeat the above four steps.	* A *mixed anhydride* is very susceptible to nucleophilic attack; azide is an excellent nucleophile, and adds with loss of CO_2 and ethoxide. * An acyl azide can rearrange; donation of the nitrogen lone pair, *1,2-alkyl shift* and loss of nitrogen gas occurs to give a highly reactive *isocyanate*; addition of benzyl alcohol to the carbonyl then occurs.
Recognise that this is not the final product, but is closely related to it.	Proton shift then gives the benzyloxy carbamate product.
Write down the structure of the final product.	

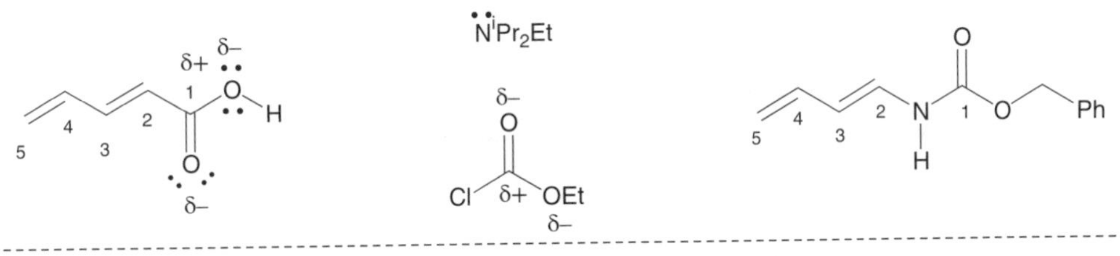

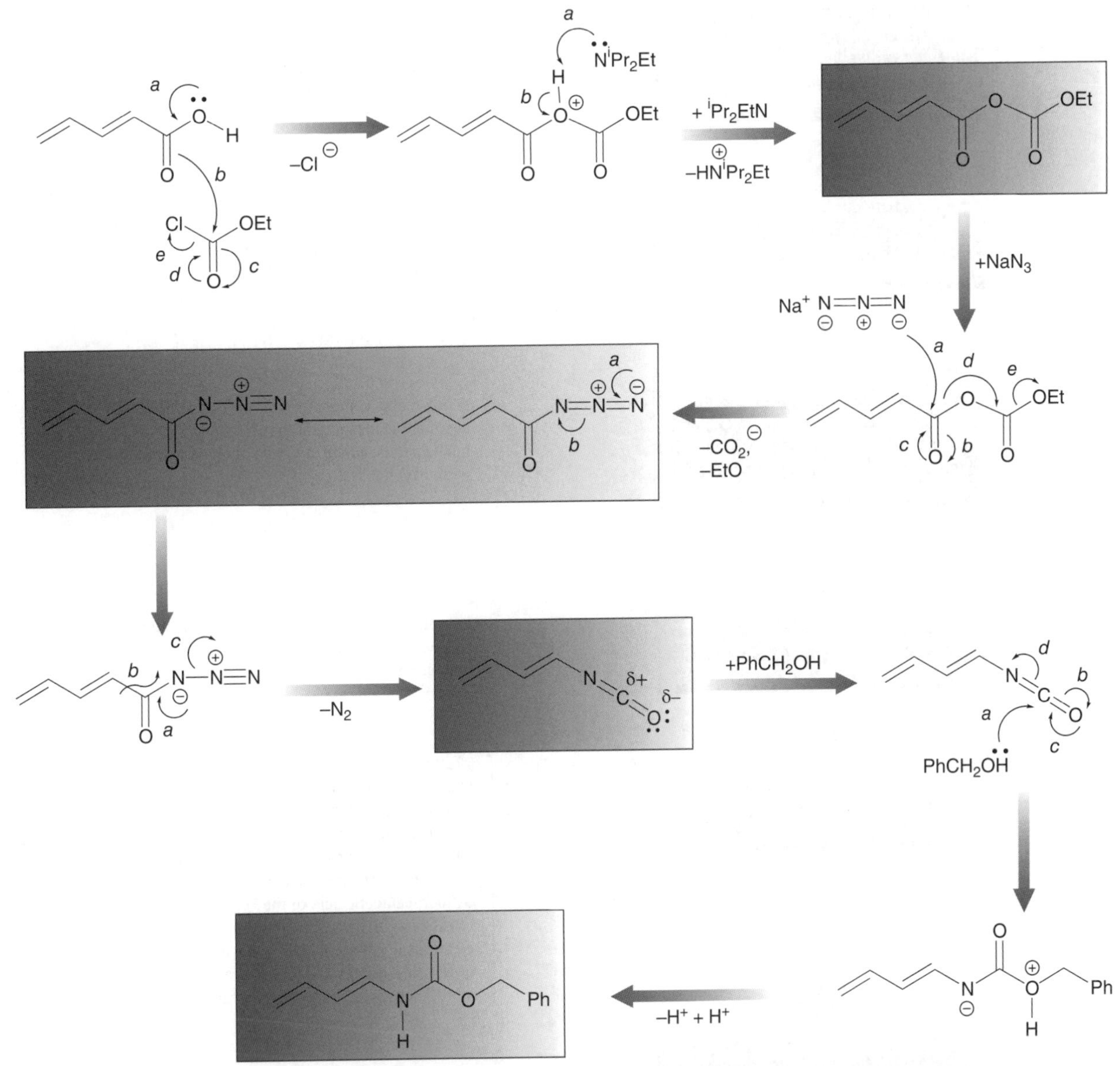

$\delta+$ $\delta-$
$\overset{\delta+}{}$ O—H
1
2
5 4 3
O
$\delta-$
$\delta-$

N^iPr_2Et

$\delta-$
O
Cl $\overset{\delta+}{}$ OEt
$\delta-$

O
5 4 3 2 N 1 O Ph
H

a

N^iPr_2Et

a
H
b O$^{\oplus}$ OEt

$+ ^iPr_2EtN$
$-Cl^{\ominus}$
$-HN^iPr_2Et$

O OEt
O O

$+NaN_3$

Na^+ N≡N≡N
$\ominus$ $\oplus$ $\ominus$
a
d e
O OEt
c b O O

$-CO_2$,
$-EtO$

N$\overset{\oplus}{=}$N≡N
$\ominus$
O

N$\overset{\oplus}{=}$N=N $\ominus$
b
O
a

c
b N$\overset{\oplus}{=}$N≡N
O $\ominus$
a

$-N_2$

N=C=O $\delta+$
$\delta-$

$+PhCH_2OH$

d
N=C=O b
a c
$PhCH_2\overset{..}{O}H$

O$^{\oplus}$
N O Ph
$\ominus$ H

$-H^+ + H^+$

O
N O Ph
H

Summary: This is an example of the Curtius rearrangement:

O
R Cl
$\xrightarrow{NaN_3}$ RNH_2

Now try questions 6.13 and 6.19

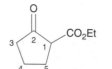

(i) NaH, then BuI
(ii) NaOH
(iii) HCl then Δ′
(iv) CF₃CO₃H

Label electrophilic/nucleophilic and acidic/basic sites of all reactants, and number identical atoms in the starting material and product.

The α-protons of β-dicarbonyl compounds are highly acidic (pK$_a$ about 13). Sodium hydride is a strong base. Iodobutane is a good electrophile, since iodide is a good leaving group and its electronegativity polarises the carbon–iodine bond.

Identify the most reactive sites, if more than one exists.

The α-position of the β-dicarbonyl enolate is the most nucleophilic (here labelled C-1) and iodobutane is the only electrophile.

Recall the characteristic reactions of the most reactive functional groups; and by considering the reaction conditions, decide which is the most appropriate.

β-Dicarbonyl compounds on reaction with good bases are readily deprotonated, giving a nucleophilic enolate, and may then be alkylated with alkyl halides.

Work through the mechanism leading to the intermediate product.

Deprotonation of the β-dicarbonyl compound is followed by *nucleophilic substitution* of iodobutane.

Repeat the above four steps.

* Ester *hydrolysis*, mediated by sodium hydroxide, gives the carboxylic acid product.
* Under acidic conditions, this is followed by decarboxylation, giving an enol intermediate, which then *tautomerises* to the ketone product.
* Addition of the hydroxyl group of trifluoroperacetic acid to the ketone gives a tetrahedral intermediate. Donation of the electrons in the H–O bond, *1,2-migration* of one of the alkyl groups and loss of the carboxylate anion then gives the product lactone.

Recognise that this is not the final product, but is closely related to it.

Not needed here.

Write down the structure of the final product.

Summary: This is an example of the Baeyer-Villiger reaction:

$$RCOR' \xrightarrow{R''CO_3H} RCO_2R'$$

(i) Mg, then CO₂ , then acidic work-up
(ii) SOCl₂ , py

(iii) CH₂N₂, Et₃N
(iv) Ag₂O, Et₃N, EtOH, Δ′

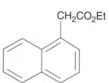

Label electrophilic/nucleophilic and acidic/basic sites of all reactants, and number identical atoms in the starting material and product.

Alkyl halides and magnesium react to form Grignard reagents, which are good carbon nucleophiles. The carbon of carbon dioxide is highly electrophilic due to the attached electronegative oxygen atoms. Magnesium is a reducing metal, and can donate two electrons.

Identify the most reactive sites, if more than one exists.

The carbon-bromine bond is the only reactive one under these conditions, and magnesium an excellent reducing agent. Carbon dioxide is the only available electrophile.

Recall the characteristic reactions of the most reactive functional groups; and by considering the reaction conditions, decide which is the most appropriate.

Magnesium and an aryl bromide react to give a *Grignard* reagent; this is nucleophilic on carbon, and reacts with carbonyl groups in a *nucleophilic addition* reaction.

Work through the mechanism leading to the intermediate product.

Nucleophilic addition of the Grignard reagant to carbon dioxide gives a carboxylate anion; acidic work-up gives the carboxylic acid product.

Repeat the above four steps.

* Thionyl chloride is electrophilic, and susceptible to attack by the carboxylic acid; a sequence of *addition-elimination* reactions gives the corresponding acid chloride.
* Diazomethane is a good nucleophile (readily apparent from one of the canonical structures), and undergoes an *addition-elimination* reaction with the acid chloride. The intermediate rearranges; in this case, departure of the *leaving group* (nitrogen) gives a carbene, which immediately undergoes a *1,2-alkyl shift*, to give a ketene. The ketene is highly electrophilic at the central carbon and is intercepted by ethanol to give the ester product.

Recognise that this is not the final product, but is closely related to it.

Tautomerisation then gives the ester product.

Write down the structure of the final product.

$$\text{RCOCl} \xrightarrow[\text{(ii) Ag}_2\text{O, R'OH}]{\text{(i) CH}_2\text{N}_2} \text{RCH}_2\text{CO}_2\text{R'}$$

Now try questions 6.15 and 6.21

6.8

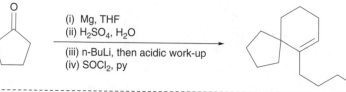

$\qquad$ (i) Mg, THF
$\qquad$ (ii) H_2SO_4, H_2O
$\qquad$ (iii) n-BuLi, then acidic work-up
$\qquad$ (iv) $SOCl_2$, py

Label electrophilic/nucleophilic and acidic/basic sites of all reactants, and number identical atoms in the starting material and product.	The carbonyl group is polarised by the *electronegativity* difference between carbon and oxygen, making the carbon electrophilic. Magnesium is a reducing metal, and can donate two electrons, in two separate single electron processes.
Identify the most reactive sites, if more than one exists.	The ketone is the only reactive electrophilic site, and magnesium the only reducing metal.
Recall the characteristic reactions of the most reactive functional groups; and by considering the reaction conditions, decide which is the most appropriate.	Reducing metals (such as Mg) can couple ketones to give 1,2-diols; this is called the Pinacol reaction.
Work through the mechanism leading to the intermediate product.	Magnesium donates a single electron to the oxygen of the C=O bond, to form a stable O–Mg bond, and a carbon centred *radical*; this occurs twice to give a magnesium bis(alkoxide). The radicals then *dimerise*, and the bisalkoxide product is protonated on work-up to give a diol.
Repeat the above four steps.	* Alcohols are basic and nucleophilic. Treatment with strong acid (H_2SO_4) protonates one of the hydroxyl groups, and this induces a *rearrangement*: the lone pair of the other hydroxyl pushes in, forces a 1,2-methyl group migration of the methyl group *anti* to the leaving group, followed by departure of the *leaving group* (H_2O).
Recognise that this is not the final product, but is closely related to it.	* Butyllithium is a carbanion source. The carbonyl group is polarised by the *electronegativity* difference between carbon and oxygen, making the carbon electrophilic. *Nucleophilic addition* of the carbanion generates an alkoxide anion.
Write down the structure of the final product.	* Thionyl chloride is highly electrophilic, and converts the alcohol to a chloride *via* an *addition–elimination* process, which then eliminates to the alkene product.

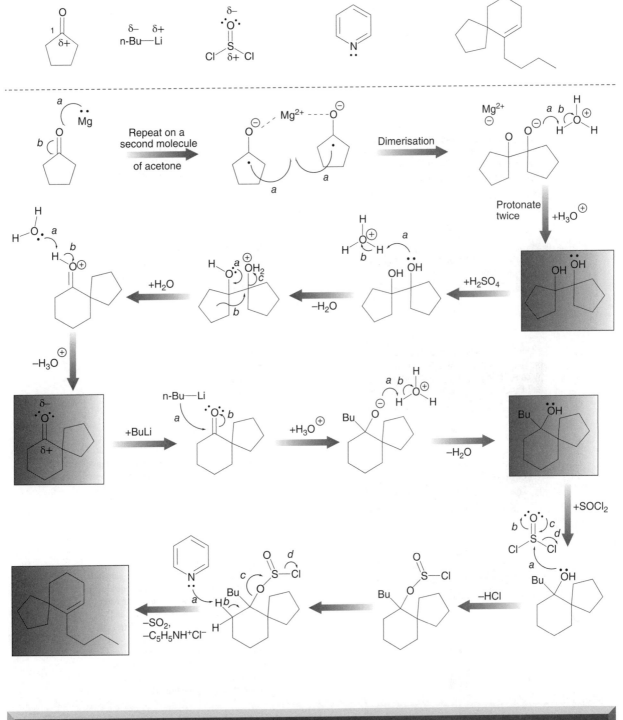

Summary: This is an example of the Pinacol-Pinacolone rearrangement:

6.9

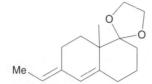

Me

(i) OsO$_4$, NaHSO$_4$, H$_2$O
(ii) TsCl, py
(iii) LiClO$_4$, CaCO$_3$

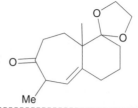

O

Me

Label electrophilic/nucleophilic and acidic/basic sites of all reactants, and number identical atoms in the starting material and product.

Alkenes are nucleophiles by virtue of the Π electron density. Osmium tetroxide is a good oxidant and the osmium is highly electrophilic.

Identify the most reactive sites, if more than one exists.

The alkenes are the only nucleophiles, with the less hindered terminal alkene better than the internal one, and osmium tetroxide is an electrophile.

Recall the characteristic reactions of the most reactive functional groups; and by considering the reaction conditions, decide which is the most appropriate.

Alkenes are very susceptible to *electrophilic addition* reactions, which with osmium tetroxide gives the product of 1,2-dihydroxylation.

Work through the mechanism leading to the intermediate product.

Alkenes react with osmium peroxide to give a cyclic osmate ester, in which the oxygens are necessarily *cis*- related. After hydrolysis under mildly acidic conditions, the 1,2-diol is released.

Repeat the above four steps.

* The less hindered secondary alcohol undergoes a *nucleophilic addition-elimination* reaction at the sulfonyl chloride, with loss of chloride; this generates a sulfonate ester.
* Lithium cations are Lewis acidic, and co-ordinate to oxygen; this activates the tosyl group to departure, and simultaneous donation of the electrons in the H-O bond, *1,2-migration* of the vinyl group and loss of the tosylate anion then gives the product.

Recognise that this is not the final product, but is closely related to it.

Not needed here.

Write down the structure of the final product.

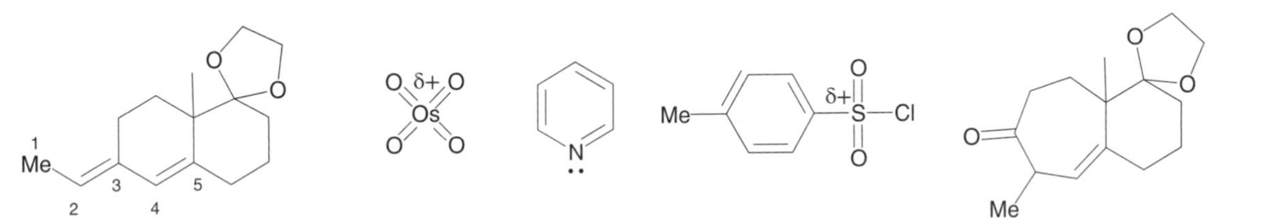

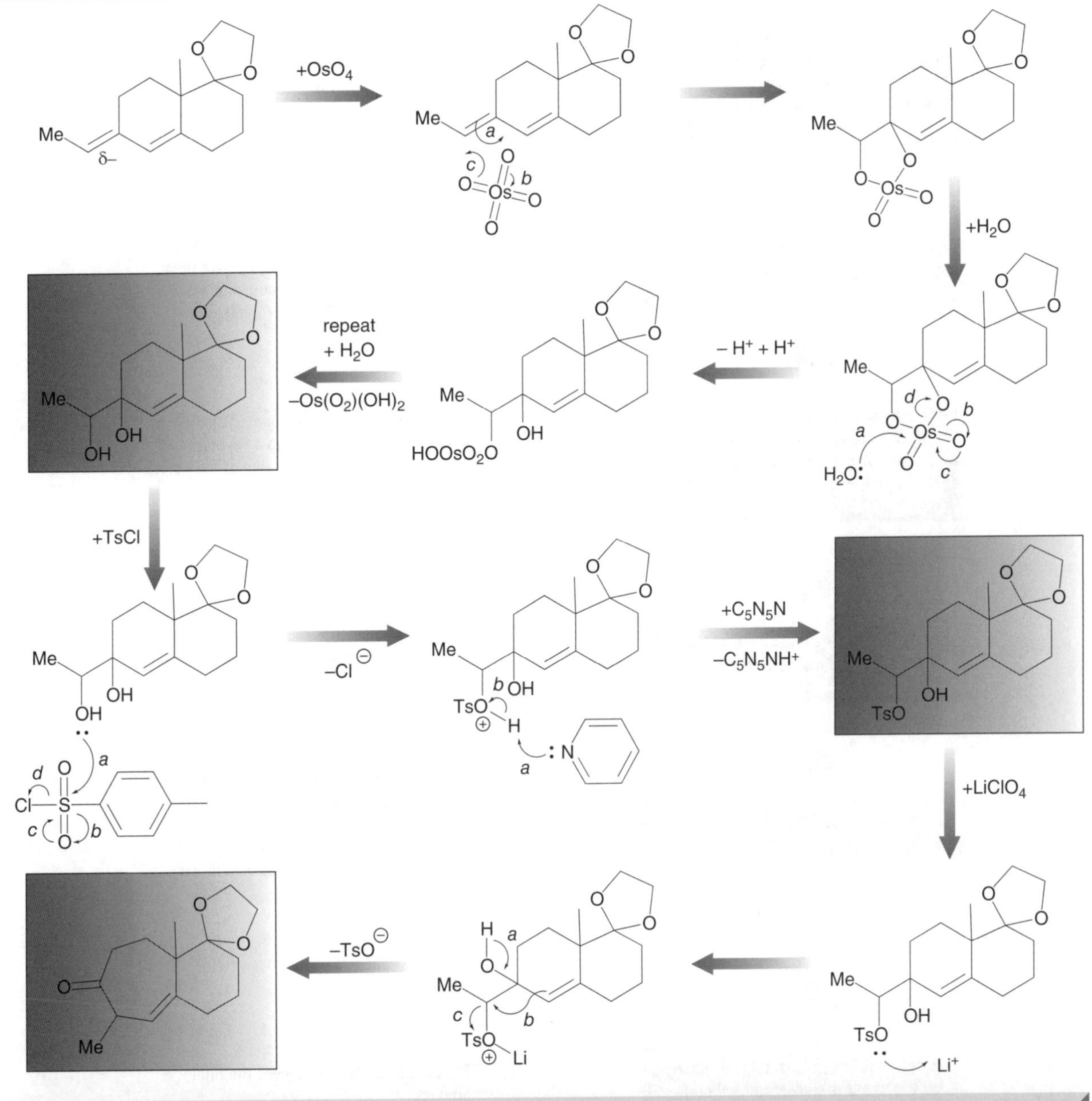

Summary: This is an example of a rrarrangement of a diol:

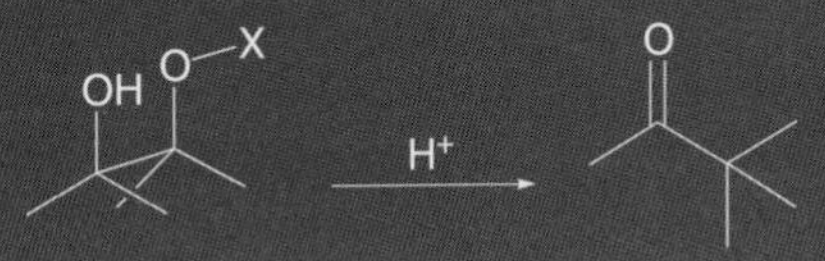

6.10

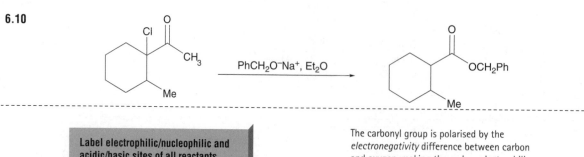

Label electrophilic/nucleophilic and acidic/basic sites of all reactants, and number identical atoms in the starting material and product.	The carbonyl group is polarised by the *electronegativity* difference between carbon and oxygen, making the carbon electrophilic. The protons α- to the ketone are acidic, and sodium benzyloxide is a good base. Chloride is good leaving group.
Identify the most reactive sites, if more than one exists.	Under basic conditions, the protons α- to the ketone are acidic, and sodium benzyloxide is a good enough base to easily remove them. Chloride is good leaving group, and easily attacked by a suitable nucleophile.
Recall the characteristic reactions of the most reactive functional groups; and by considering the reaction conditions, decide which is the most appropriate.	Ketones are readily deprotonated at the α-position and will undergo a Favorskii rearrangement if the other side of the ketone is suitably activated.
Work through the mechanism leading to the intermediate product.	The α-position of the enolate attacks the other side to displace chloride leaving group, to give a cyclopropanone.
Repeat the above four steps.	Cyclopropanones are particularly susceptible to *nucleophilic addition* due to *ring strain*. Sodium benzyloxide is also a good nucleophile, and undergoes an *addition-elimination* reaction at the carbonyl with concomitant ring opening, to generate a stabilised enolate product.
Recognise that this is not the final product, but is closely related to it.	Protonation on work-up gives the final product.
Write down the structure of the final product.	

Cl
O
$\delta+$
1
CH_3
2
Me

$PhCH_2O^{\ominus}\ Na^+$

Me
O
1
OCH_2Ph
2
Me

- -

Cl
O
CH_3
Me

$+PhCH_2O^-$

Cl
O
a
b
H
$^{\ominus}OCH_2Ph$
H
H
Me

$-HOCH_2Ph$

Cl
O
$^{\ominus}$
CH_2
Me

b
Cl
O
CH_2
a
$^{\ominus}$
Me

$-Cl^{\ominus}$

O
Me

$+PhCH_2O^-$

b
O
a
c
$^{\ominus}OCH_2Ph$
Me

O
$^{\ominus}$
OCH_2Ph
Me

$+H_3O^{\oplus}$

H
$O^{\oplus}$
H
H
a
b
$^{\ominus}$
O
OCH_2Ph
Me

$-H_2O$

O
OCH_2Ph
Me

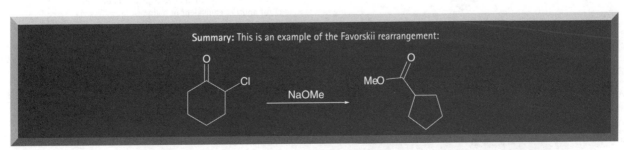

Summary: This is an example of the Favorskii rearrangement:

O
Cl

$\xrightarrow{\text{NaOMe}}$

MeO
O

Now try question 6.16

181

6.11

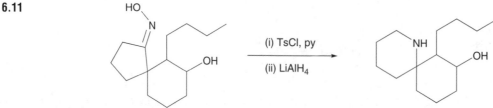

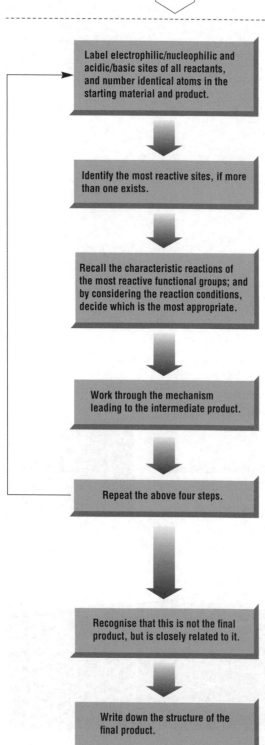

Label electrophilic/nucleophilic and acidic/basic sites of all reactants, and number identical atoms in the starting material and product.	Oximes are good nucleophiles (oxygen possesses lone pairs). *p*-TsCl is an acid chloride derived from a sulfonic acid, and is electrophilic at the sulfur atom. Pyridine is a mild base.
Identify the most reactive sites, if more than one exists.	The alcohol function is the most nucleophilic site and the sulfonyl chloride is the most electrophilic site.
Recall the characteristic reactions of the most reactive functional groups; and by considering the reaction conditions, decide which is the most appropriate.	Reaction of a sulfonyl acid chloride with the alcohol nucleophile gives an ester product, in this case a sulfonate, with expulsion of a chloride leaving group.
Work through the mechanism leading to the intermediate product.	The alcohol undergoes a *nucleophilic addition–elimination* reaction at the sulfonyl group, with loss of chloride; this generates a sulfonate ester.
Repeat the above four steps.	* Under the conditions of the reaction, this tosylate may rearrange, to generate a carbocation intermediate, which is intercepted by water to give an enol. Tautomerisation leads to formation of the amide product. * The amide product is readily reduced by LiAlH₄, by nucleophilic addition of hydride anion (H⁻), to give an amine product. This proceeds by initial addition of hydride, elimination of lithium hydroxide, and addition of another equivalent of hydride.
Recognise that this is not the final product, but is closely related to it.	Protonation on work-up gives the amine product
Write down the structure of the final product.	

Summary: This is an example of the Beckmann rearrangement:

Now try question 6.17

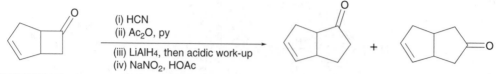

| Label electrophilic/nucleophilic and acidic/basic sites of all reactants, and number identical atoms in the starting material and product. | The carbonyl group is polarised by the *electronegativity* difference between carbon and oxygen, making the carbon electrophilic. Cyanide is a weak acid ($pK_a = 9$), which ionises to give a *cyanide* anion, which is nucleophilic. |

| Identify the most reactive sites, if more than one exists. | The ketone is the only electrophilic site and cyanide is the only nucleophile. |

| Recall the characteristic reactions of the most reactive functional groups; and by considering the reaction conditions, decide which is the most appropriate. | The carbon of the ketone is susceptible to *nucleophilic addition*, which with cyanide leads to formation of a cyanohydrin. |

| Work through the mechanism leading to the intermediate product. | *Nucleophilic addition* of the cyanide anion to the carbonyl group gives an alkoxide; protonation on work-up gives the corresponding alcohol. |

| Repeat the above four steps. | * Esterification of the alcohol with acetic anhydride gives an ester product, resulting from an *addition–elimination* reaction.
* Reduction of the nitrile group proceeds by *nucleophilic addition* of hydride twice to give an amine.
* *Nitrosation* of the amine group with nitrous acid gives a diazo group, which is an excellent leaving group. This departs, with concomitant migration either of the two adjacent bonds, to give the two products. |

| Recognise that this is not the final product, but is closely related to it. | Not needed here. |

| Write down the structure of the final product. | |

Summary: This is an example of the Tiffenau–Demjanov Rearrangement:

Now try question 6.18

6.13

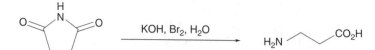

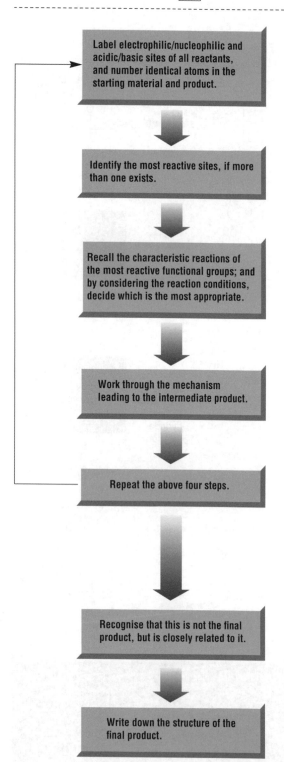

Phthalimide is acidic, since its conjugate base is resonance stabilised. Bromine is an electrophile, since it is electronegative and bromide is a good leaving group. Sodium hydroxide is a good base.

The phthalimide group is the most acidic site, and hydroxide the strongest base. Bromine is the only electrophile.

Amides react with halogens under basic conditions to give the *N*-bromoamide product, which will readily rearrange under basic conditions.

Phthalimide is readily deprotonated by hydroxide, to give the corresponding anion; this reacts with bromine to give the *N*-bromophthalimide product.

* Under the conditions of this reaction, the *N*-bromophthalimide product rearranges with loss of bromide, giving an isocyanate after a 1,2-carbon shift. This is intercepted by water to give the carbamic acid, which spontaneously decarboxylates under basic conditions.
* Amide hydrolysis under basic conditions proceeds *via* an *addition–elimination* process, to give the ring opened amino acid product.

Not needed here.

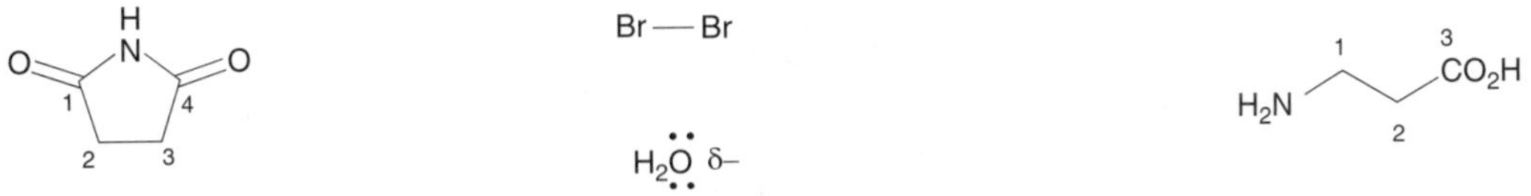

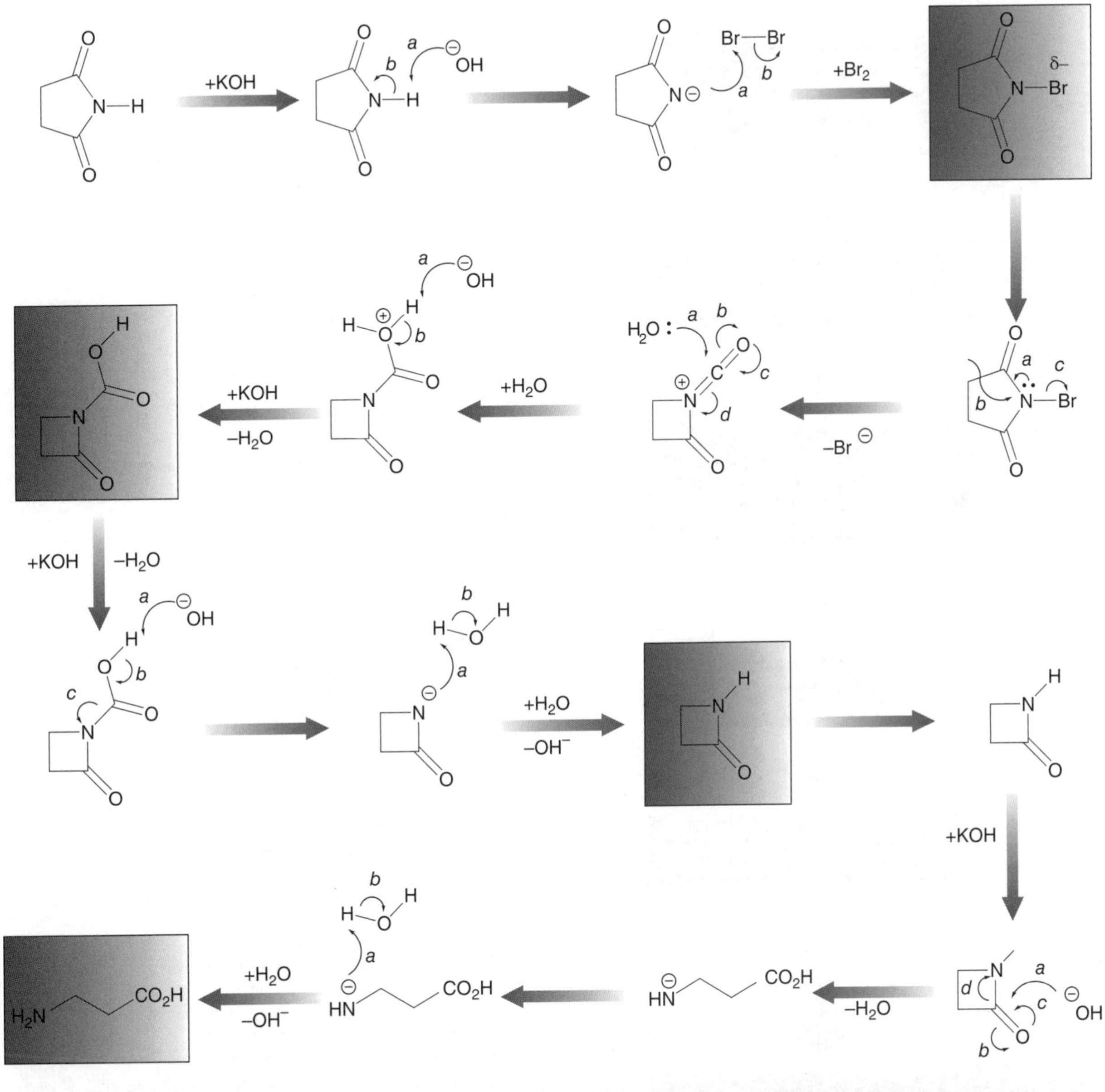

Br — Br

H_2O δ−

H_2N —1— —2—3 CO_2H

+KOH

+Br2

Br—Br

δ−

N—Br

−Br⊖

H_2O

+H2O

−H2O

+KOH

+KOH   −H2O

+H2O

−OH⁻

+KOH

−H2O

+H2O

−OH⁻

H_2N CO_2H

Summary: This is an example of the Hofmann rearrangement:

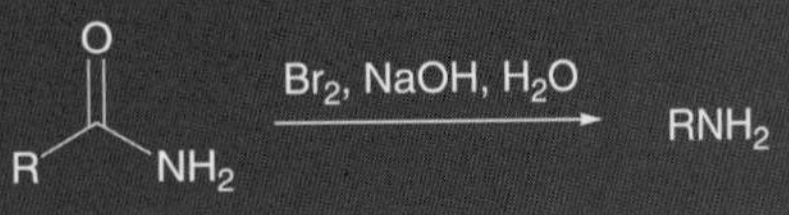

$$\text{R}-\text{C(=O)}-\text{NH}_2 \xrightarrow{\ \text{Br}_2,\ \text{NaOH},\ \text{H}_2\text{O}\ } \text{RNH}_2$$

Now try question 6.19

6.14

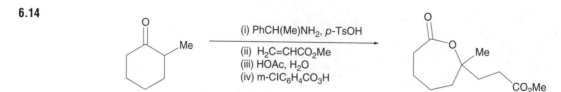

$$(i) \ PhCH(Me)NH_2, \ p\text{-}TsOH$$
$$(ii) \ H_2C=CHCO_2Me$$
$$(iii) \ HOAc, \ H_2O$$
$$(iv) \ m\text{-}ClC_6H_4CO_3H$$

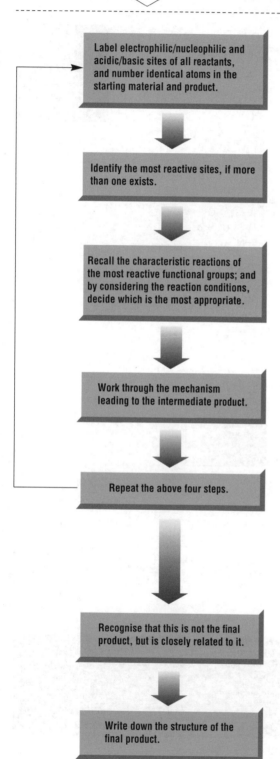

Label electrophilic/nucleophilic and acidic/basic sites of all reactants, and number identical atoms in the starting material and product.

Identify the most reactive sites, if more than one exists.

Recall the characteristic reactions of the most reactive functional groups; and by considering the reaction conditions, decide which is the most appropriate.

Work through the mechanism leading to the intermediate product.

Repeat the above four steps.

Recognise that this is not the final product, but is closely related to it.

Write down the structure of the final product.

The carbonyl group is polarised by the *electronegativity* difference between carbon and oxygen, making the carbon electrophilic. The α-protons of a carbonyl group are acidified by *inductive* withdrawl, and the resulting enolate is resonance stabilised and a good nucleophile. However, although the amine is a good nucleophile, it is not basic enough to deprotonate a ketone. Toluenesulfonic acid is a good acid.

The ketone is the only electrophilic site, and the amine the most nucleophilic reagent. Toluenesulfonic acid is a good acid.

Ketones and secondary amines readily react to give enamines under acidic catalysis.

The ketone is protonated, and *nucleophilic addition* of the amine to the ketone then proceeds, and in this case *elimination* of the intermediate to an *enamine* is possible.

* An enamine is an excellent nucleophile (compare with an enol) at the β-carbon since nitrogen is strongly nucleophilic. Methyl acrylate is a good electrophile, and *nucleophilic addition* then occurs.
* Acid-catalysed iminium *hydrolysis* gives the ketone product.
* Addition of the hydroxyl group of trifluoroperacetic acid to the ketone gives a tetrahedral intermediate. Donation of the electrons in the H-O bond, *1,2-migration* of one of the alkyl groups and loss of the carboxylate anion then gives the product lactone.

Not needed here.

Summary: This is an example of the Baeyer–Villiger reaction:

Summary: This is an example of the Baeyer–Villiger reaction:

$$RCOR' \xrightarrow{R''CO_3H} RCO_2R'$$

Now try question 6.20

6.15

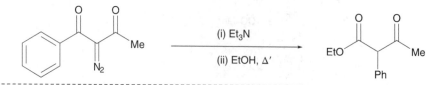

(i) Et$_3$N

(ii) EtOH, Δ'

Flow chart steps	Explanatory notes
Label electrophilic/nucleophilic and acidic/basic sites of all reactants, and number identical atoms in the starting material and product.	The α-protons of β-dicarbonyl compounds are highly acidic (pK$_a$ about 13) and give an enolate which is a good nucleophile. Tosyl azide is a very good electrophile, and triethylamine is a base.
Identify the most reactive sites, if more than one exists.	The α-position of the β-dicarbonyl compound is the most acidic, and the corresponding enolate is the most nucleophilic reagent. The terminal nitrogen of the azide is the most electrophilic site.
Recall the characteristic reactions of the most reactive functional groups; and by considering the reaction conditions, decide which is the most appropriate.	Enolates will react with electrophiles; in this case the electrophile is an unusual azide, which itself possesses a leaving group.
Work through the mechanism leading to the intermediate product.	*Addition* of the enolate to the azide acceptor is followed by elimination of tosylamide, to give the diazo product (two canonical forms can be written).
Repeat the above four steps.	Thermal collapse with loss of nitrogen gas gives a carbene, which rearranges to give a ketene, which is electrophilic on the central carbon atome. This reacts with ethanol to give the ester product.
Recognise that this is not the final product, but is closely related to it.	Not needed here.
Write down the structure of the final product.	

190

EtO̤H

Summary: This is an example of the Wolff rearrangement:

RXH
X = O, NH

Now try question 6.21

7 Ligand coupling processes

The coupling of two organic compounds, an organic halide (RX) and an organometallic (R'M), using transition metal catalysts to give products of type R–R' is synthetically very powerful, having found wide application in research and industry.

$$R\!-\!X \quad + \quad R'\!-\!M \quad \xrightarrow{\text{catalyst}} \quad R\!-\!R' \quad + \quad M\!-\!X$$

R and R' are usually sp^2-hybridised, but R' may also be sp^3-or sp-hybridised
M = Sn, B, Zn, Mg, Si, Zr
X = I, Br, Cl, OSO$_2$CF$_3$
catalyst = Pd(0) and appropriate ligands

It is driven by a metal-centred redox process, beginning with oxidative addition of the halide substrate to the metal, ligand exchange and then reductive elimination leading to product formation as shown below:

$$M(0)L_2 \quad \xrightarrow[\;]{\substack{\text{Oxidative}\\\text{addition}\\+X-Y}} \quad Y\!-\!\overset{X}{\underset{}{M(II)L_2}} \quad \xrightarrow[+RM',\,-Y]{\text{Ligand}\atop\text{exchange}} \quad R\!-\!\overset{X}{\underset{}{M(II)L_2}} \quad \xrightarrow[\substack{\text{(ligand}\\\text{coupling)}}]{\substack{\text{Reductive}\\\text{elimination}}} \quad X\text{-}R + M(0)L_2$$

This process is most widely exemplified for palladium(0) catalysts, for which a number of named reactions have been developed, which vary according to the identity of the organometallic coupling partner, including:

Stille coupling, using organostannanes (M = SnR''$_3$)
Hiyama coupling, using organosilanes (M = SiR''$_3$)
Suzuki coupling, using organoborons (M = B(OH)$_3$). The Suzuki–Miyaura B-alkyl variant permits C(sp^3)–C(sp^2) coupling, using an alkylborane as the coupling partner
Molander coupling, using organofluoroboronates (M = BF$_4^-$)
Negishi coupling, using organozincs (M = ZnR''$_3$), and Kumada coupling, using organomagnesiums (Grignards) (M = MgR'').

The value in these processes comes from the fact that the starting materials are generally readily available, and many functional groups on the R and R' reacting partners are tolerated (and include CO$_2$R, CN, OH, CHO, NO$_2$), and that the reactions may be accelerated using microwave irradiation. Mechanistically, these processes are quite different to the ones we have seen in the earlier chapters of this book, and proceed in a cycle in which the metal species is regenerated (and hence is catalytic), involving a sequence of oxidative addition, transmetallation and reductive elimination steps. Oxidative addition of the palladium catalyst into the R–X bond initially gives a *cis*-complex that rapidly isomerises to its *trans*-isomer (see below), and importantly, both the oxidative addition and reductive elimination steps occur with retention of configuration.

$$R\!-\!X \quad \xrightarrow{Pd(0)L_2} \quad R\!-\!\overset{\overset{L}{|}}{\underset{\underset{X}{|}}{Pd}}\!-\!L \quad \xrightarrow{\text{fast}} \quad R\!-\!\overset{\overset{L}{|}}{\underset{\underset{L}{|}}{Pd}}\!-\!X$$

$$\qquad\qquad\qquad\qquad\qquad\qquad \textit{cis-} \qquad\qquad\qquad\qquad \textit{trans-}$$

How to Solve Organic Reaction Mechanisms: A Stepwise Approach, First Edition. Mark G. Moloney.
© 2015 John Wiley & Sons, Ltd. Published 2015 by John Wiley & Sons, Ltd. Companion website: www.wiley.com/go/moloney/mechanisms

Strongly σ-donating ligands, such as trialkylphosphines and *N*-heterocyclic carbenes, increase electron density at the metal, and this accelerates the oxidative addition step, and since this appears to be the rate determining step, ligands can have profound effects on the course of these coupling reactions; however, this is a highly specialist topic and will not be considered in any detail here. This species then reacts with the organometallic derivative in a transmetallation process, with the following relative order of ligand transfer from organostannanes:

$$\text{organo} = \text{alkynyl} > \text{alkenyl} > \text{aryl} > \text{allyl} = \text{benzyl} > \alpha\text{-alkoxyalkyl} > \text{alkyl}$$

Reductive elimination then generates the coupled product, R–R′, and releases the reduced metal Pd(0), which is ready to catalyse the next cycle of the process.

Two important variants in this process replace the organometallic partner with an alkyne or alkene nucleophile. These reactions are also catalysed by Pd(0) in a redox-driven reaction cycle but differ in some of the key details of the reaction cycle.

Sonogashira coupling of organohalides with acetylenes is a process which is normally assisted by the presence of a copper(I) catalyst, in which a copper(I) acetylide is generated *in situ*, and it is this transient organometallic derivative that delivers one of the 'R' groups:

Sonogashira coupling $\quad$ R—X $\quad + \quad$ ≡—R′ $\quad \xrightarrow[\text{base}]{\text{Pd(0),Cu(I),}} \quad$ R——≡——R′ $\quad + \quad$ M—X

Heck coupling of organohalides with alkenes follows a different pathway to the palladium-catalysed couplings outlined earlier, because in this case, the organopalladium intermediate undergoes a carbopalladation process followed by a β-hydride elimination under thermodynamically controlled conditions, leading to the formation of an (*E*)-alkene product.

Heck coupling $\quad$ R—X $\quad + \quad$ ═／R′ $\quad \xrightarrow{\text{Pd(0)}} \quad$ R／═＼R′ $\quad + \quad$ M—X

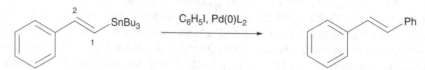

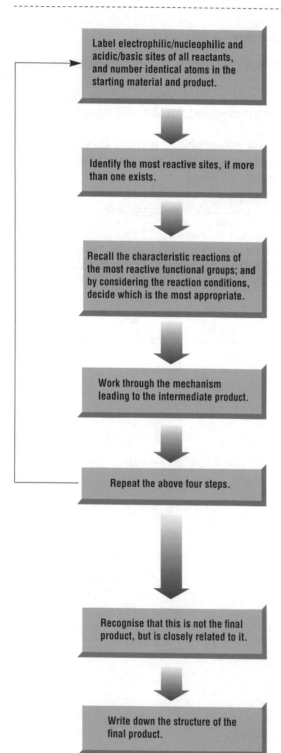

Pd(0) is readily oxidised to Pd(II), and a polarised C–I bond is capable of accepting the released electrons, to form a Pd–C bond.

In this case, the only reactive sites are the C–I bond and the Pd(0) metal species.

Palladium(0) mediates the coupling of organostannanes and organoiodides in the Stille reaction.

Organohalides (usually with Csp^2–X bonds) and Pd(0) readily participate in ligand coupling processes. Pd(0) is readily oxidised to Pd(II), with simultaneous oxidative addition into the C–I bond, to give a *cis*-organopalladium complex, which rapidly equilibrates to the *trans*-system. This *trans*–complex undergoes transmetalation (metal-metal exchange) with the vinyl stannane to give another palladium complex, with the two principal ligands in a *trans*-arrangement. Equilibration to the *cis*-system permits reductive elimination, giving the product by ligand coupling, and releases $Pd(0)L_2$ which then catalyses the next reaction.

Not needed here.

Summary: This is an example of the Stille reaction which couples an organostannane and a organohalide under palladium catalysis:

$$R'\text{-}I + RSnBu_3 \xrightarrow{Pd(0)} R'\text{-}R$$

Now try questions 7.8 and 7.9

7.2

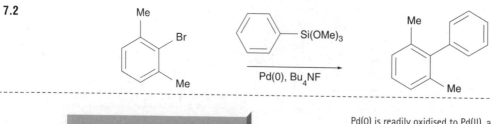

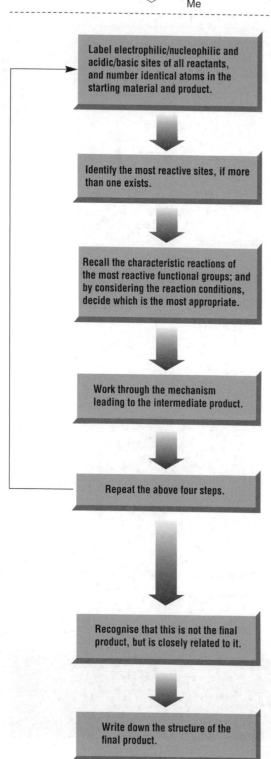

Pd(0) is readily oxidised to Pd(II), and a polarised C–Br bond is capable of accepting the released electrons, to form a Pd–C bond. Fluoride forms strong Si–F bonds, and addition to the Si centre activates the organic ligand to transfer by transmetallation.

In this case, the only reactive sites are the C–Br bond and the Pd(0) metal species.

Palladium(0) mediates the coupling of organosilanes and organoiodides in the Hiyama reaction.

Organohalides (usually with Csp^2–X bonds) and Pd(0) readily participate in ligand coupling processes. Pd(0) is readily oxidised to Pd(II), with simultaneous oxidative addition into the C–Br bond, to give a *cis*-organopalladium complex, which rapidly equilibrates to the *trans*-system. This *trans*-complex undergoes transmetalation (metal–metal exchange) with the activated silane to give another palladium complex, with the two principal ligands in a *trans*-arrangement. Equilibration to the *cis*-system permits reductive elimination, giving the product by ligand coupling, and releases $Pd(0)L_2$ which then catalyses the next reaction.

Not needed here.

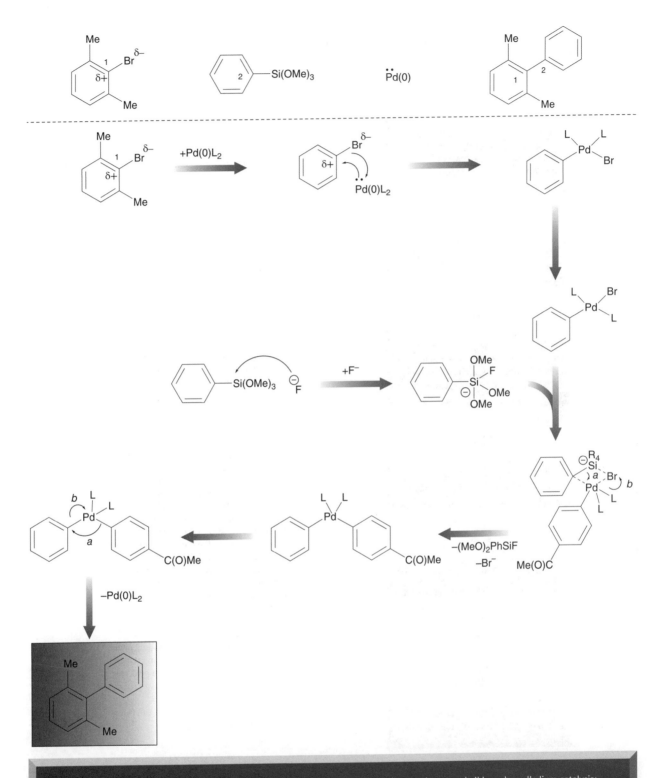

Summary: This is an example of the Hiyama reaction which couples an organosilane and a organohalide under palladium catalysis:

$$R'\!-\!I \ + \ RSiX_3 \ \xrightarrow{\ Pd(0),\ Bu_4NF\ } \ R'\!-\!R$$

Now try questions 7.10 and 7.16

7.3

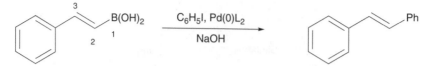

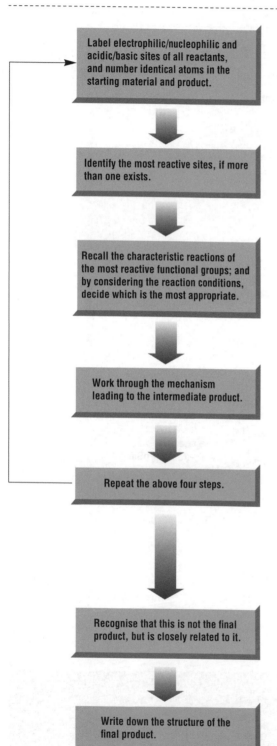

Label electrophilic/nucleophilic and acidic/basic sites of all reactants, and number identical atoms in the starting material and product.

Pd(0) is readily oxidised to Pd(II), and a polarised C–I bond is capable of accepting the released electrons, to form a Pd–C bond.

Identify the most reactive sites, if more than one exists.

In this case, the only reactive sites are the C–I bond and the Pd(0) metal species.

Recall the characteristic reactions of the most reactive functional groups; and by considering the reaction conditions, decide which is the most appropriate.

Palladium(0) mediates the coupling of organoboranes and organoiodides in the Suzuki reaction.

Work through the mechanism leading to the intermediate product.

Repeat the above four steps.

Organohalides (usually with Csp^2–X bonds) and Pd(0) readily participate in ligand coupling processes. Pd(0) is readily oxidised to Pd(II), with simultaneous oxidative addition into the C–I bond, to give a *cis*-organopalladium complex, which rapidly equilibrates to the *trans*-system. Hydroxide adds to the boronic acid to make an ate complex, which is activated to ligand exchange. The *trans*-palladium complex undergoes transmetalation (metal–metal exchange) with the vinyl boronate ate complex to give another palladium complex, with the two principal ligands in a *trans*-arrangement. Equilibration to the *cis*-system permits reductive elimination, giving the product by ligand coupling, and releases Pd(0) L_2 which then catalyses the next reaction.

Recognise that this is not the final product, but is closely related to it.

Not needed here.

Write down the structure of the final product.

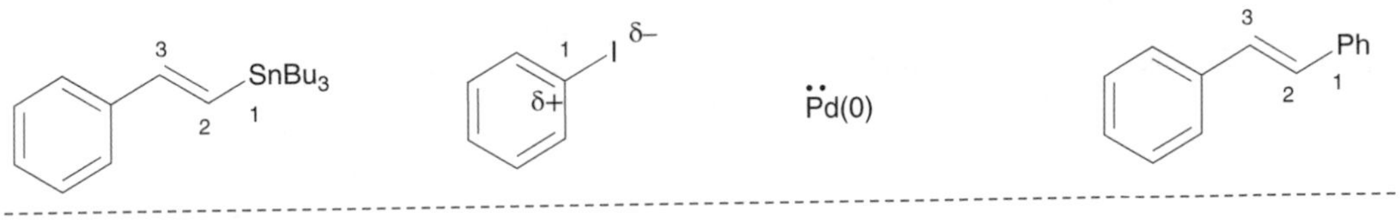

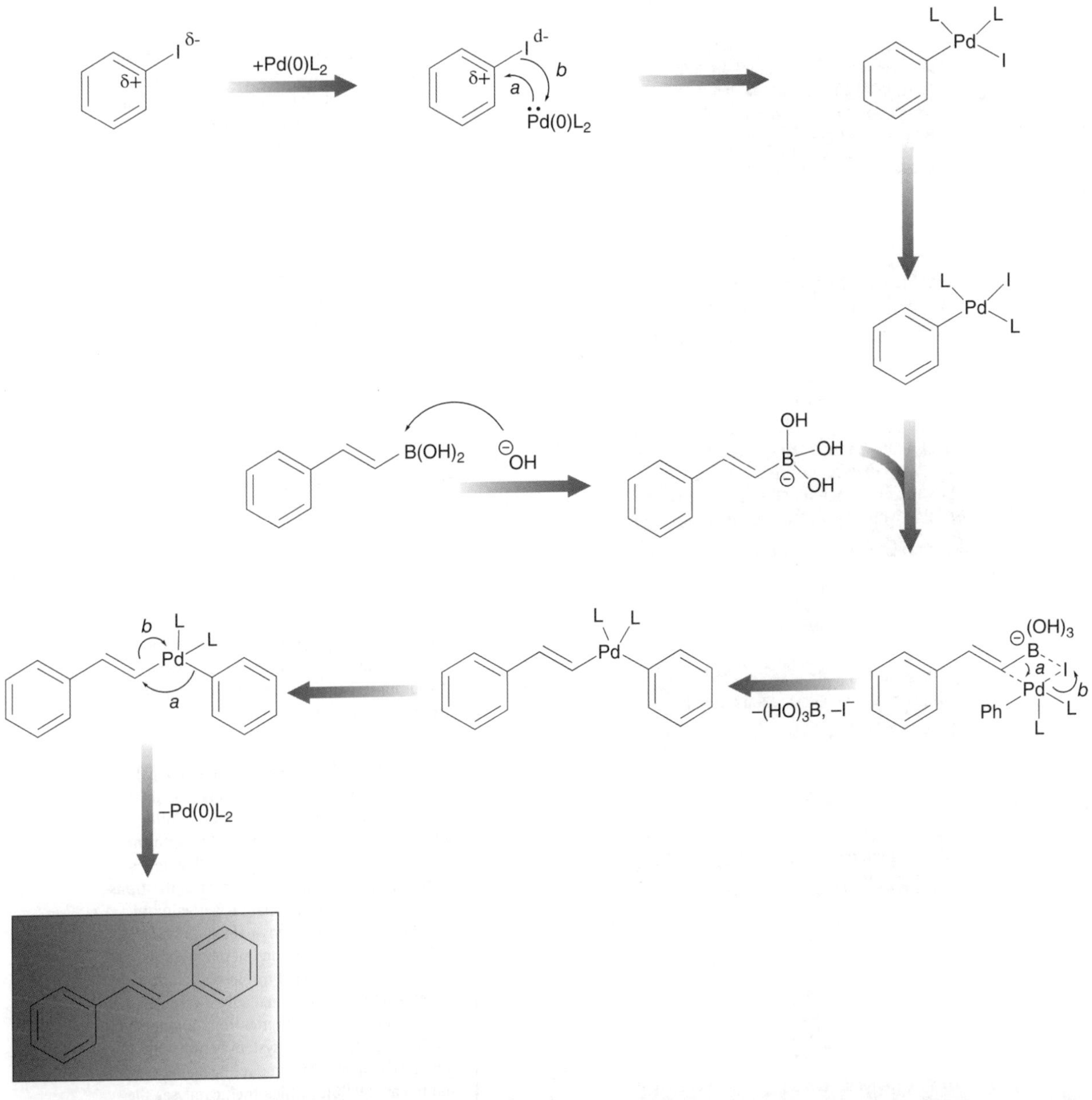

SnBu$_3$

Ph

+Pd(0)L$_2$

L Pd L / I

L Pd I / L

B(OH)$_2$ $^{\ominus}$OH

OH B OH OH $^{\ominus}$

$^{\ominus}$(OH)$_3$ B Pd Ph L I L

L Pd L

L Pd L

–(HO)$_3$B, –I$^-$

–Pd(0)L$_2$

Summary: This is an example of the Suzuki reaction which couples an organoboronic acid and a organohalide under palladium catalysis:

$$R'-I \; + \; RB(OH)_2 \xrightarrow{\text{Pd(0)}} R'-R$$

Now try questions 7.11 and 7.17

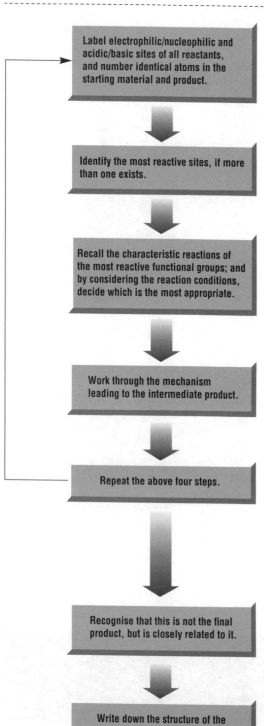

Phenols are nucleophilic, and a sulfonic anhydride is electrophilic with a good leaving group (trifluoromethansulfonate, OTf). Pd(0) is readily oxidised to Pd(II), and a polarised C-OTf bond is capable of accepting the released electrons, to form a Pd–C bond.

In this case, the only reactive sites are the phenol, the acid anhydride, the C-OTf bond and the Pd(0) metal species.

Phenols, like alcohols, react with acid anhydrides to form esters.

The phenol is a nucleophile which reacts with the electrophilic acid anhydride, by a process of addition-elimination, leading to expulsion of the good leaving group, triflate.

Organo triflates (usually with Csp^2-X bonds) and Pd(0) readily participate in ligand coupling processes. Pd(0) is readily oxidised to Pd(II), with simultaneous oxidative addition into the C-OTf bond, to give a *cis*-organopalladium complex, which rapidly equilibrates to the *trans*-system. The trifluorosulfonate is activated to ligand exchange. The *trans*-palladium complex undergoes transmetalation (metal–metal exchange) with the vinyl trifluoroboronate complex to give another palladium complex, with the two principal ligands in a *trans*-arrangement. Equilibration to the *cis*- system permits reductive elimination, giving the product by ligand coupling, and releases $Pd(0)L_2$ which then catalyses the next reaction.

Not needed here.

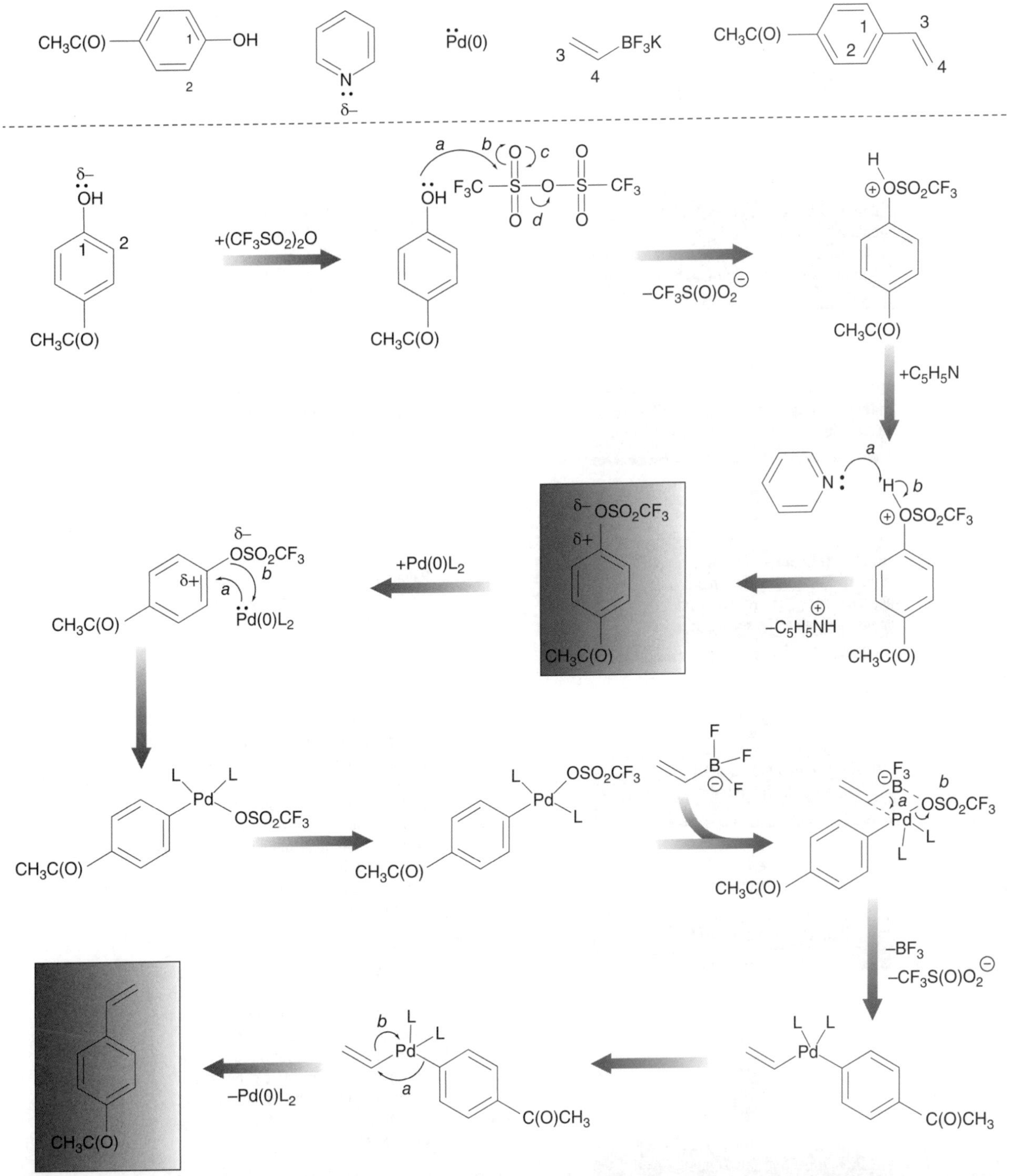

CH3C(O)
OH
Pd(0)
BF3K
δ–
+(CF3SO2)2O
F3C
CF3
–CF3S(O)O2
⊕OSO2CF3
+C5H5N
–C5H5NH
OSO2CF3
δ+
+Pd(0)L2
Pd(0)L2
Pd
L
–BF3
C(O)CH3
–Pd(0)L2

Summary: This is an example of the Molander reaction which couples an organotrifluoroborate and a organohalide under palladium catalysis:

$$R'{-}I + RBF_3K \xrightarrow{Pd(0)} R'{-}R$$

Now try questions 7.12 and 7.18

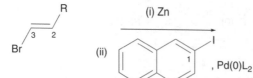

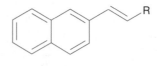

(i) Zn

(ii) , Pd(0)L₂

Label electrophilic/nucleophilic and acidic/basic sites of all reactants, and number identical atoms in the starting material and product.

Zn metal is able to oxidatively insert into C–X bonds. Pd(0) is readily oxidised to Pd(II), and a polarised C–I bond is capable of accepting the released electrons, to form a Pd–C bond.

Identify the most reactive sites, if more than one exists.

In this case, the only reactive sites are the C–I and C–Br bonds in each of the starting materials, and the Zn(0) and Pd(0) metal species.

Recall the characteristic reactions of the most reactive functional groups; and by considering the reaction conditions, decide which is the most appropriate.

Palladium(0) mediates the coupling of organozincs and organoiodides in the Negishi reaction.

Work through the mechanism leading to the intermediate product.

Organohalides (usually with Csp^2–X bonds) and Pd(0) readily participate in ligand coupling processes. Pd(0) is readily oxidised to Pd(II), with simultaneous oxidative addition into the C–I bond, to give *cis*–organopalladium complex, which rapidly equilibrates to the *trans*–system. The bromoalkene reacts with zinc metal by oxidative addition, to give an organozinc species, which is activated to ligand exchange. The *trans*-palladium complex undergoes transmetalation (metal-metal exchange) with the organozinc species to give another palladium complex, with the two principal ligands in a *trans*-arrangement. Equilibration to the *cis*-system permits reductive elimination, giving the product by ligand coupling, and releases Pd(0) which then catalyses the next reaction.

Repeat the above four steps.

Recognise that this is not the final product, but is closely related to it.

Not needed here.

Write down the structure of the final product.

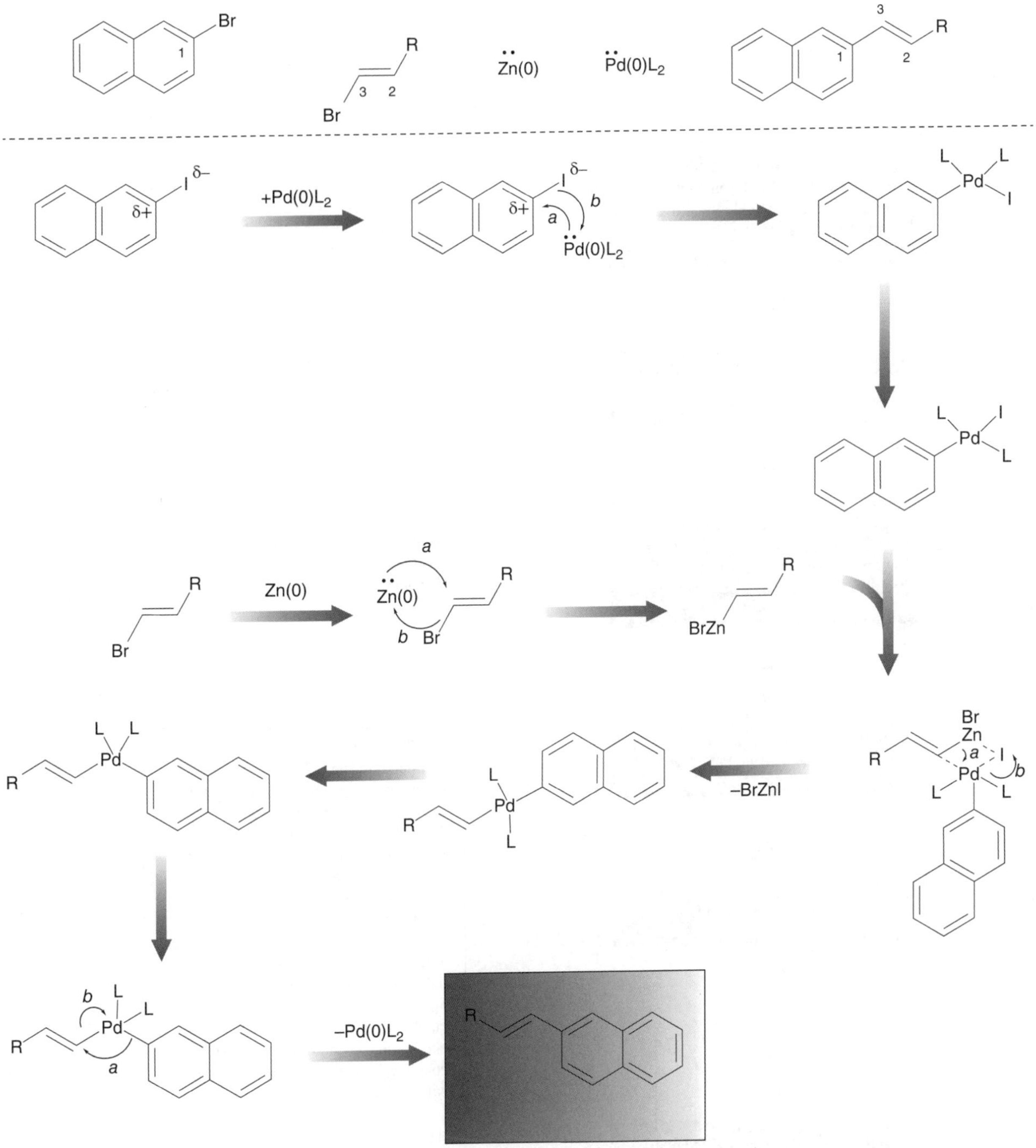

Br

R

3 2
Br

$\ddot{Zn}(0)$

$\ddot{Pd}(0)L_2$

3
R
1
2

$\delta-$
I
$\delta+$

+Pd(0)L$_2$

$\delta-$
I
b
a
$\delta+$
$\ddot{Pd}(0)L_2$

L   L
Pd
I

L   I
Pd
L

R

Zn(0)

a
$\ddot{Zn}(0)$
R
b Br

R

BrZn

L   L
Pd
R

−BrZnI

L
Pd
R
L

R

Br
Zn
a I
Pd b
L   L

L   L
b Pd
R
a

−Pd(0)L$_2$

R

Summary: This is an example of the Negishi reaction which couples an alkenylzinc and an organohalide under palladium catalysis:

$$R' - I \ + \ RCH=CHBr \xrightarrow{Pd(0)} RCH=CHR'$$

Now try questions 7.13 and 7.19

7.6

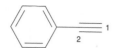 

$$\xrightarrow{\begin{array}{c}C_6H_5I,\ Pd(0)L_2\\ Et_3N,\ CuI\end{array}}$$

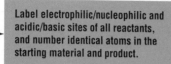

Label electrophilic/nucleophilic and acidic/basic sites of all reactants, and number identical atoms in the starting material and product.

Pd(0) is readily oxidised to Pd(II), and a polarised C–I bond is capable of accepting the released electrons, to form a Pd–C bond. The acetylene function is acidic, and readily deprotonated.

Identify the most reactive sites, if more than one exists.

In this case, the only reactive sites are the C–I bond, the alkyne and the Pd(0) metal species.

Recall the characteristic reactions of the most reactive functional groups; and by considering the reaction conditions, decide which is the most appropriate.

Palladium(0) mediates the coupling of acetylenes and organoiodides in the Sonagashira reaction.

Work through the mechanism leading to the intermediate product.

Organohalides (usually with Csp^2–X bonds) and Pd(0) readily participate in ligand coupling processes. Pd(0) is readily oxidised to Pd(II), with simultaneous oxidative addition into the C–I bond, to give a *cis*-organopalladium complex, which rapidly equilibrates to the *trans*–system.
Triethylamine base deprotonates the alkyne, to give an acetylide anion, which reacts with the copper iodide to give a copper acetylide, which is activated to ligand exchange.
The *trans*-palladium complex undergoes transmetalation (metal–metal exchange) with the copper acetylide to give another palladium complex, with the two principal ligands in a *trans*-arrangement.
Equilibration to the *cis*–system permits reductive elimination, giving the product by ligand coupling, and releases $Pd(0)L_2$ which then catalyses the next reaction.

Repeat the above four steps.

Recognise that this is not the final product, but is closely related to it.

Not needed here.

Write down the structure of the final product.

Summary: This is an example of the Sonagashira reaction which couples an alkyne and an organohalide under palladium catalysis:

$$R'\text{-}I + RC \equiv CH \xrightarrow{\text{Pd(0)}} RC \equiv CR'$$

Now try questions 7.14 and 7.20

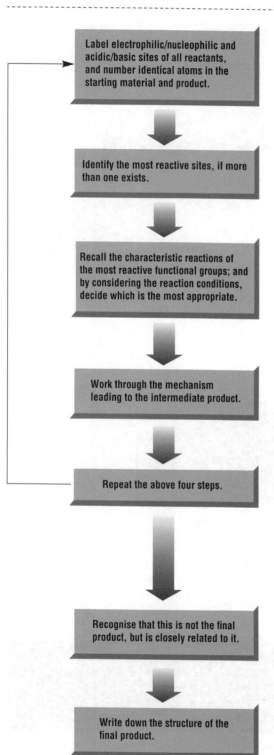

Label electrophilic/nucleophilic and acidic/basic sites of all reactants, and number identical atoms in the starting material and product.

Pd(0) is readily oxidised to Pd(II), and a polarised C–I bond is capable of accepting the released electrons, to form a Pd–C bond.

Identify the most reactive sites, if more than one exists.

In this case, the only reactive sites are the C–I bond, the alkene and the Pd(0) metal species.

Recall the characteristic reactions of the most reactive functional groups; and by considering the reaction conditions, decide which is the most appropriate.

Palladium(0) mediates the coupling of alkenes and organoiodides in the Heck reaction.

Work through the mechanism leading to the intermediate product.

Organohalides (usually with Csp^2–X bonds) and Pd(0) readily participate in ligand coupling processes. Pd(0) is readily oxidised to Pd(II), with simultaneous oxidative addition into the C–I bond, to give a *cis*-organopalladium complex, which rapidly equilibrates to the *trans*–system. The *trans*-palladium complex undergoes addition to the carbon-carbon double bond, in a process called *carbopalladation*, to give a new organopalladium species. β-Elimination of a palladium hydride gives the product, and releases [Pd(0)HIL$_2$] which eliminates HI to give Pd(0)L$_2$ which then catalyses the next reaction.

Repeat the above four steps.

Recognise that this is not the final product, but is closely related to it.

Not needed here.

Write down the structure of the final product.

Summary: This is an example of the Heck reaction which couples an alkene and an organohalide under palladium catalysis.

$$R'\text{-}I \ + \ RCH{=}CH_2 \ \xrightarrow{\ Pd(0)\ } \ RCH{=}CHR'$$

Now try questions 7.15 and 7.21

7.8

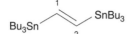

Bu₃Sn —1— SnBu₃
         ‖
         2

$$\xrightarrow[\text{(ii) }C_6H_5I,\ Pd(0)L_2]{\text{(i) MeLi then ClC(O)OEt}}$$

CO₂Me

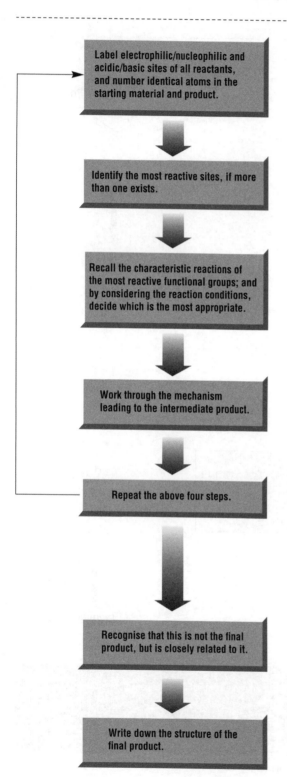

Label electrophilic/nucleophilic and acidic/basic sites of all reactants, and number identical atoms in the starting material and product.

MeLi is a very strong base but also an excellent nucleophile. Ethylchloroformate is an excellent electrophile, with chloride being a good leaving group. Pd(0) is readily oxidised to Pd(II), and a polarised C–Br bond is capable of accepting the released electrons, to form a Pd–C bond. Chlorosilanes are weakly electrophilic.

Identify the most reactive sites, if more than one exists.

In this case, reactive sites are the vinyl stannane, the C–I bond and the Pd(0) metal species.

Recall the characteristic reactions of the most reactive functional groups; and by considering the reaction conditions, decide which is the most appropriate.

MeLi is able to displace tin from the vinylic system, and the vinylic carbanion so generated is nucleophilic and reacts with the chloroformate to give an ester product.

Work through the mechanism leading to the intermediate product.

MeLi attacks the organo tin, forming an organotin derivative and releasing tin from the vinylic system. The vinylic carbanion thus generated is unstable and reacts with the chloroformate to give an ester product by an addition-elimination process.

Repeat the above four steps.

* Organohalides (usually with Csp²–X bonds) and Pd(0) readily participate in ligand coupling processes. Pd(0) is readily oxidised to Pd(II), with simultaneous oxidative addition into the C–I bond, to give a *cis*-organopalladium complex, which rapidly equilibrates to the *trans*-system. this *trans*-complex undergoes transmetalation (metal-metal exchange) with the vinyl stannane to give another palladium complex, with the two principal ligands in a *trans*-arrangement.
* Equilibration to the *cis*-system permits reductive elimination, giving the product by ligand coupling, and releases Pd(0)L₂ which then catalyses the next reaction.

Recognise that this is not the final product, but is closely related to it.

Not needed here.

Write down the structure of the final product.

Now try question 7.9

7.9

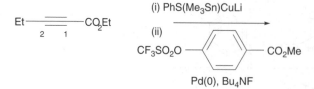

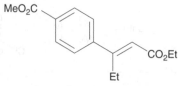

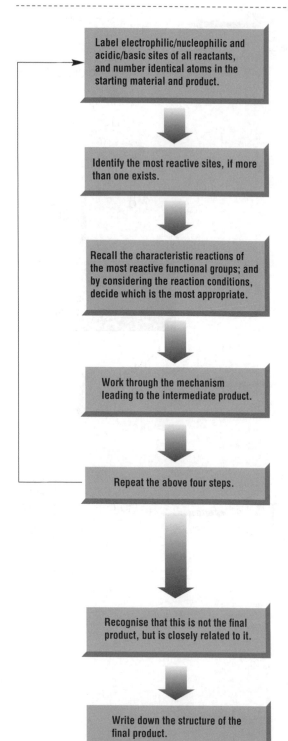

The cuprate species is an excellent nucleophile, and capable of delivering trimethylstannyl anion to an electrophile. The alkyne is an excellent electrophile, being activated by the ester group. Pd(0) is readily oxidised to Pd(II), and a polarised C–OTf bond is capable of accepting the released electrons, to form a Pd–C bond.

In this case, reactive sites are the stannanylcuprate, the C–OTf bond and the Pd(0) metal species.

Activated alkynes react with nucleophiles by conjugate addition, and the carbanion thus generated can be quenched with electrophiles, in this case acid.

The cuprate delivers stannyl anion into the alkyne system, and the vinylic carbanion so generated reacts with the acid to give an ester product.

* Organohalides (usually with Csp^2–X bonds) and Pd(0) readily participate in ligand coupling processes. Pd(0) is readily oxidised to Pd(II), with simultaneous oxidative addition into the C–OTf bond, to give a *cis*-organopalladium complex, which rapidly equilibrates to the *trans*-system. This *trans*-complex undergoes transmetalation (metal–metal exchange) with the vinyl stannane to give another palladium complex, with the two principal ligands in a *trans*-arrangement.
* Equilibration to the *cis*-system permits reductive elimination, giving the product by ligand coupling, and releases $Pd(0)L_2$ which then catalyses the next reaction.

Not needed here.

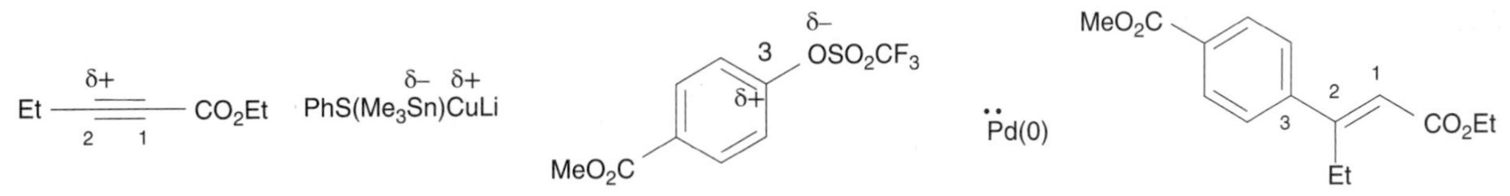

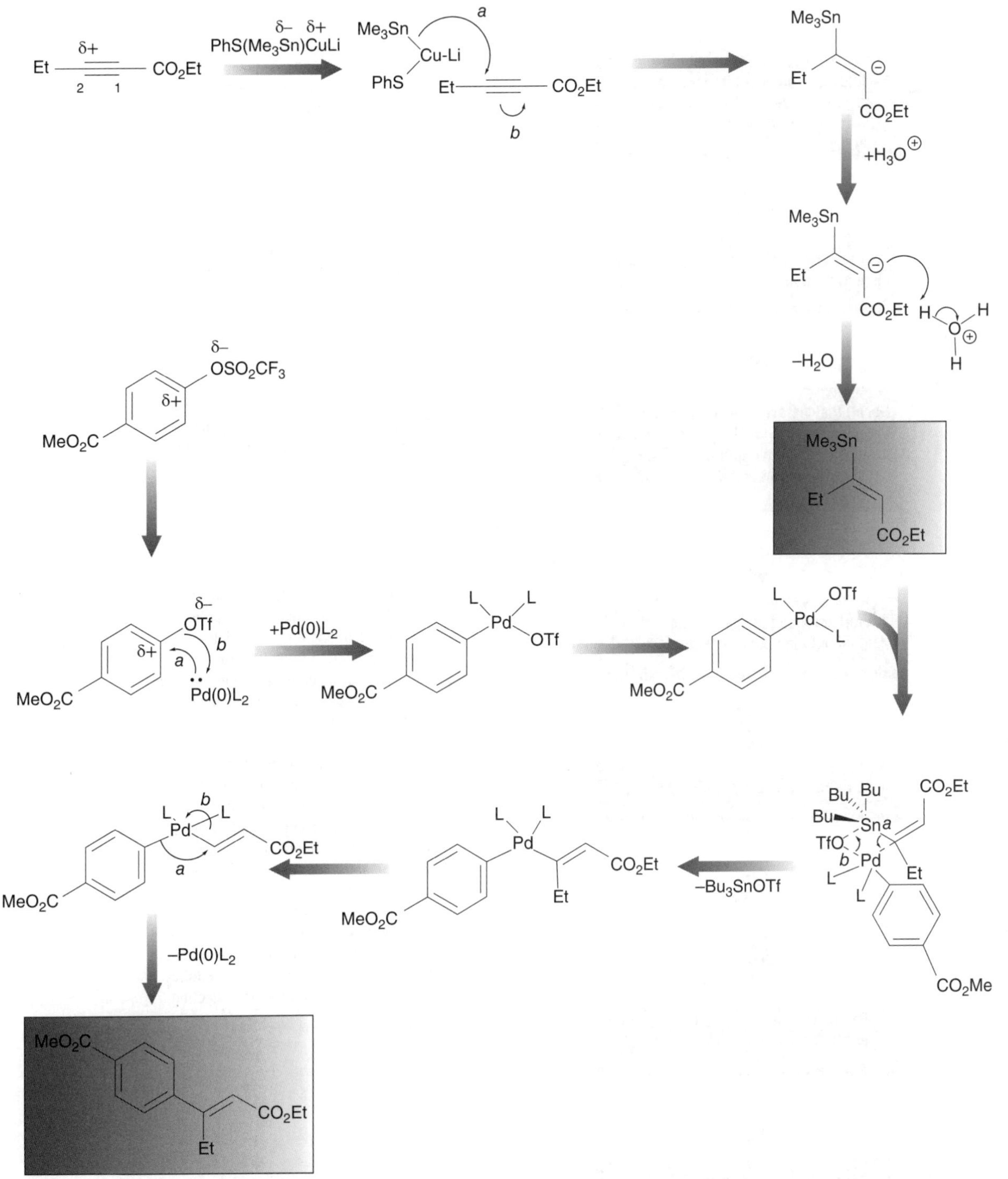

Et —≡— CO₂Et δ+ 2 1

PhS(Me₃Sn)CuLi δ− δ+

3 OSO₂CF₃ δ−
MeO₂C δ+

Pd(0)

MeO₂C — / — CO₂Et 1 2 3 Et

Et —≡— CO₂Et δ+ 2 1

PhS(Me₃Sn)CuLi δ− δ+

Me₃Sn Cu–Li a
PhS Et —≡— CO₂Et b

Me₃Sn Et ⊖ CO₂Et

+H₃O⊕

Me₃Sn Et ⊖ CO₂Et H O H ⊕ H

−H₂O

Me₃Sn Et CO₂Et

OSO₂CF₃ δ−
MeO₂C δ+

δ− OTf b
δ+ a
MeO₂C Pd(0)L₂

+Pd(0)L₂

L L Pd OTf
MeO₂C

L OTf Pd L
MeO₂C

Bu Bu CO₂Et Bu Sn a TfO b Pd Et L L CO₂Me

−Bu₃SnOTf

L L Pd CO₂Et Et MeO₂C

L b L Pd CO₂Et a MeO₂C

−Pd(0)L₂

MeO₂C — / — CO₂Et Et

Now try question 7.16

7.10

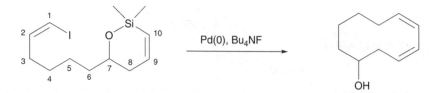

Pd(0), Bu₄NF →

Label electrophilic/nucleophilic and acidic/basic sites of all reactants, and number identical atoms in the starting material and product.

Pd(0) is readily oxidised to Pd(II), and a polarised C–Br bond is capable of accepting the released electrons, to form a Pd–C bond. Fluoride forms strong Si–F bonds, and this activates the organic ligand to transfer by transmetallation.

Identify the most reactive sites, if more than one exists.

In this case, the reactive sites are the C–I bond, the vinyl silane and the Pd(0) metal species.

Recall the characteristic reactions of the most reactive functional groups; and by considering the reaction conditions, decide which is the most appropriate.

Palladium(0) mediates the coupling of organosilanes and organoiodides in the Hiyama reaction.

Work through the mechanism leading to the intermediate product.

Organolhalides (usually with Csp²–X bonds) and Pd(0) readily participate in ligand coupling processes. Pd(0) is readily oxidised to Pd(II), with simultaneous oxidative addition into the C–I bond, to give a *cis*–organopalladium complex, which rapidly equilibrates to the *trans*–system.
Attack by fluoride on the silicon at the far end of the molecule generates and activated ate–complex. The *trans*–palladium complex undergoes transmetalation (metal–metal exchange) with the activated silane to give another palladium complex, with the two principal ligands in a *trans*–arrangement. Equilibration to the *cis*–system permits reductive elimination, giving the product by ligand coupling, and releases Pd(0)L₂ which then catalyses the next reaction.

Repeat the above four steps.

Recognise that this is not the final product, but is closely related to it.

Not needed here.

Write down the structure of the final product.

Now try question 7.16

7.11

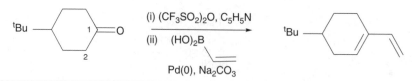

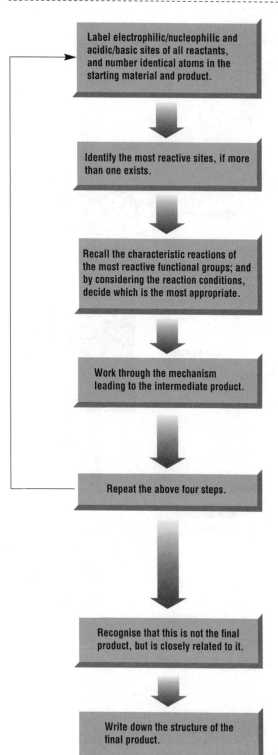

Ketones exist in equilbrium with their enol forms, which are nucleophilic, and an acid anhydride is electrophilic with a good leaving group (trifluoromethansulfonate, OTf). Pd(0) is readily oxidised to Pd(II), and a polarised C–OTf bond is capable of accepting the released electrons, to form a Pd–C bond.

In this case, only reactive sites are the ketone, the acid anhydride, the C–OTf bond and the Pd(0) metal species.

Equiibration of the keto to the enol form generates a nucleophile. Enols, like alcohols, react with acid anhydrides to form esters.

Equiibration of the keto to the enol form generates a nucleophile which reacts with the electrophilic acid anhydride, by a process of addition–elimination, leading to expulsion of the good leaving group, triflate.

Organo triflates (usually with Csp^2–X bonds)– and Pd(0) readily participate in ligand coupling processes. Pd(0) is readily oxidised to Pd(II), with simultaneous oxidative addition into the C–OTf bond, to give a *cis*–organopalladium complex, which rapidly equilibrates to the *trans*–system. Hydroxide, generated from the carbonate base, adds to the boronic acid to make an ate complex, which is activated to ligand exchange. The *trans*-palladium complex undergoes transmetalation (metal–metal exchange) with the vinyl boronate ate complex to give another palladium complex, with the two principal ligands in a *trans*-arrangement. Equilibration to the *cis*–system permits reductive elimination, giving the product by ligand coupling, and releases $Pd(0)L_2$ which then catalyses the next reaction.

Not needed here.

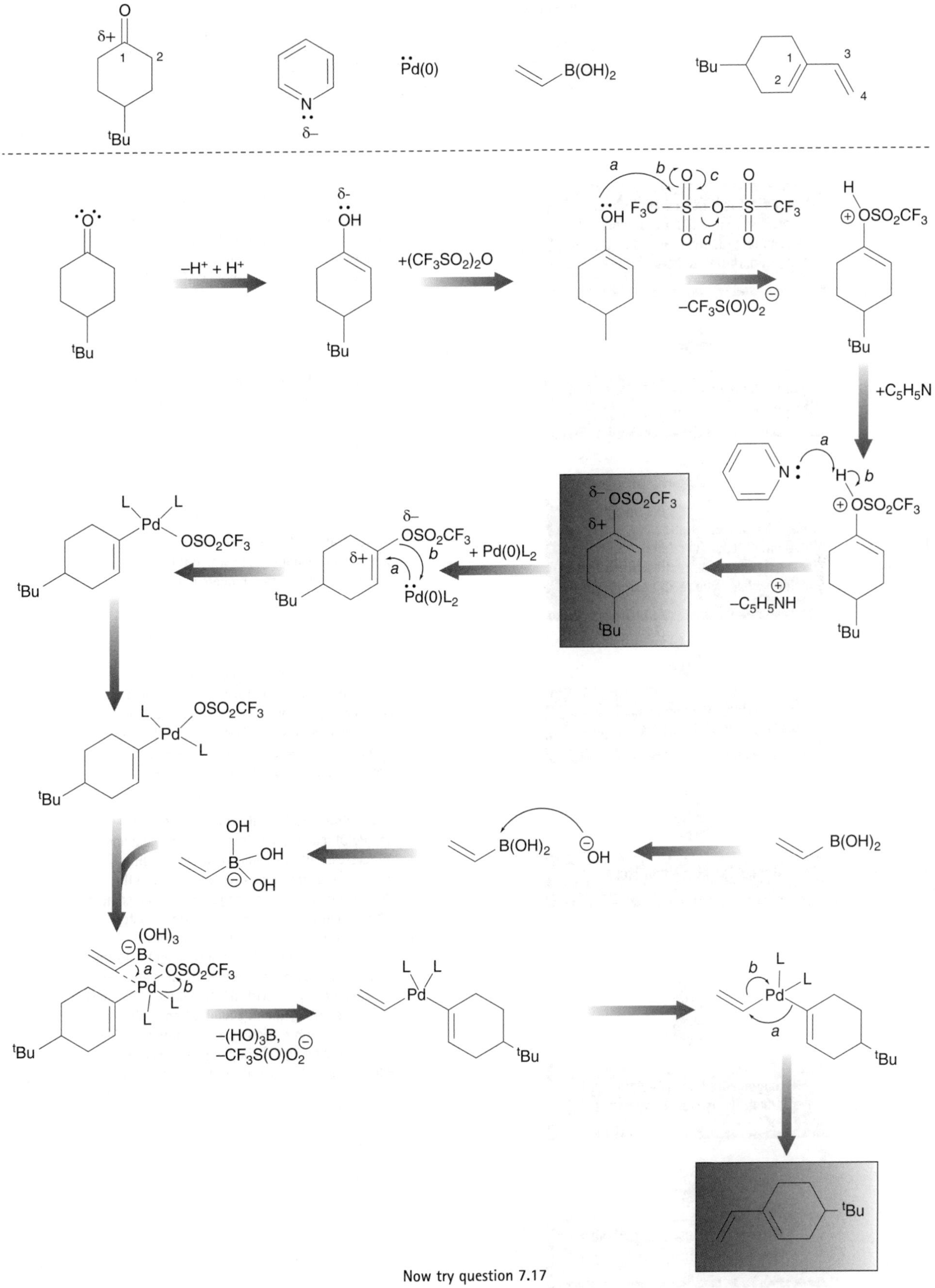

$\delta+$

1

2

^tBu

N

$\ddot{P}d(0)$

$B(OH)_2$

^tBu 1 2 3 4

$-H^+ + H^+$

$+(CF_3SO_2)_2O$

^tBu

OH

^tBu

$F_3C-S-O-S-CF_3$

$-CF_3S(O)O_2^{\ominus}$

$H-OSO_2CF_3$

^tBu

$+C_5H_5N$

$\delta- OSO_2CF_3$

$\delta+$

^tBu

$-C_5H_5NH^{\oplus}$

$\oplus OSO_2CF_3$

^tBu

$+ Pd(0)L_2$

$\ddot{P}d(0)L_2$

OSO_2CF_3

^tBu

$L \quad L$

$Pd-OSO_2CF_3$

^tBu

$L \quad OSO_2CF_3$

Pd

L

^tBu

$B(OH)_2$

OH

$B \quad OH$

OH

$^{\ominus}OH$

$B(OH)_2$

$(OH)_3$

B

a

OSO_2CF_3

Pd

b

L

L

^tBu

$-(HO)_3B,$
$-CF_3S(O)O_2^{\ominus}$

$L \quad L$

Pd

^tBu

$b \quad L$

$Pd-L$

a

^tBu

^tBu

Now try question 7.17

215

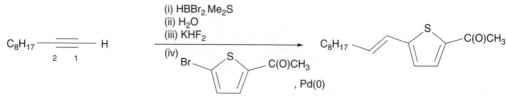

- -

Label electrophilic/nucleophilic and acidic/basic sites of all reactants, and number identical atoms in the starting material and product.

Alkynes are good nucleophiles, and dibromoborane is an excellent electrophile. Pd(0) is readily oxidised to Pd(II), and a polarised C—Br bond is capable of accepting the released electrons, to form a Pd—C bond.

Identify the most reactive sites, if more than one exists.

In this case, the only reactive sites are the alkyne, the borane, the C—Br bond and the Pd(0) metal species.

Recall the characteristic reactions of the most reactive functional groups; and by considering the reaction conditions, decide which is the most appropriate.

Hydroboration of alkynes readily generates vinylboranes, in which the boron adds to the least hindered position.

Work through the mechanism leading to the intermediate product.

Hydroboration of the alkyne readily occurs in a concerted process to generate a vinylborane, in which the boron adds to the least hindered position. Hydrolysis of the B—Br bond, and then conversion to B—F bonds, occurs by a series of SN2-like processes.

Repeat the above four steps.

Organobromides (usually with Csp²-X bonds) and Pd(0) readily participate in ligand coupling processes. Pd(0) is readily oxidised to Pd(II), with simultaneous oxidative addition into the C—Br bond, to give a *cis*-organopalladium complex, which rapidly equilibrates to the *trans*-system. The *trans*-palladium complex undergoes transmetalation (metal–metalexchange) with the vinyl trifluoroboronate complex to give another palladium complex, with the two principal ligands in a *trans*-arrangement. Equilibration to the *cis*-system permits reductive elimination, giving the product by ligand coupling, and releases Pd(0)L₂ which then catalyses the next reaction.

Recognise that this is not the final product, but is closely related to it.

Not needed here.

Write down the structure of the final product.

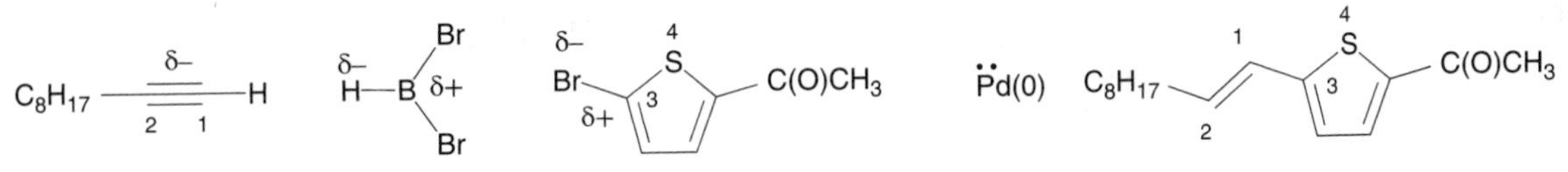

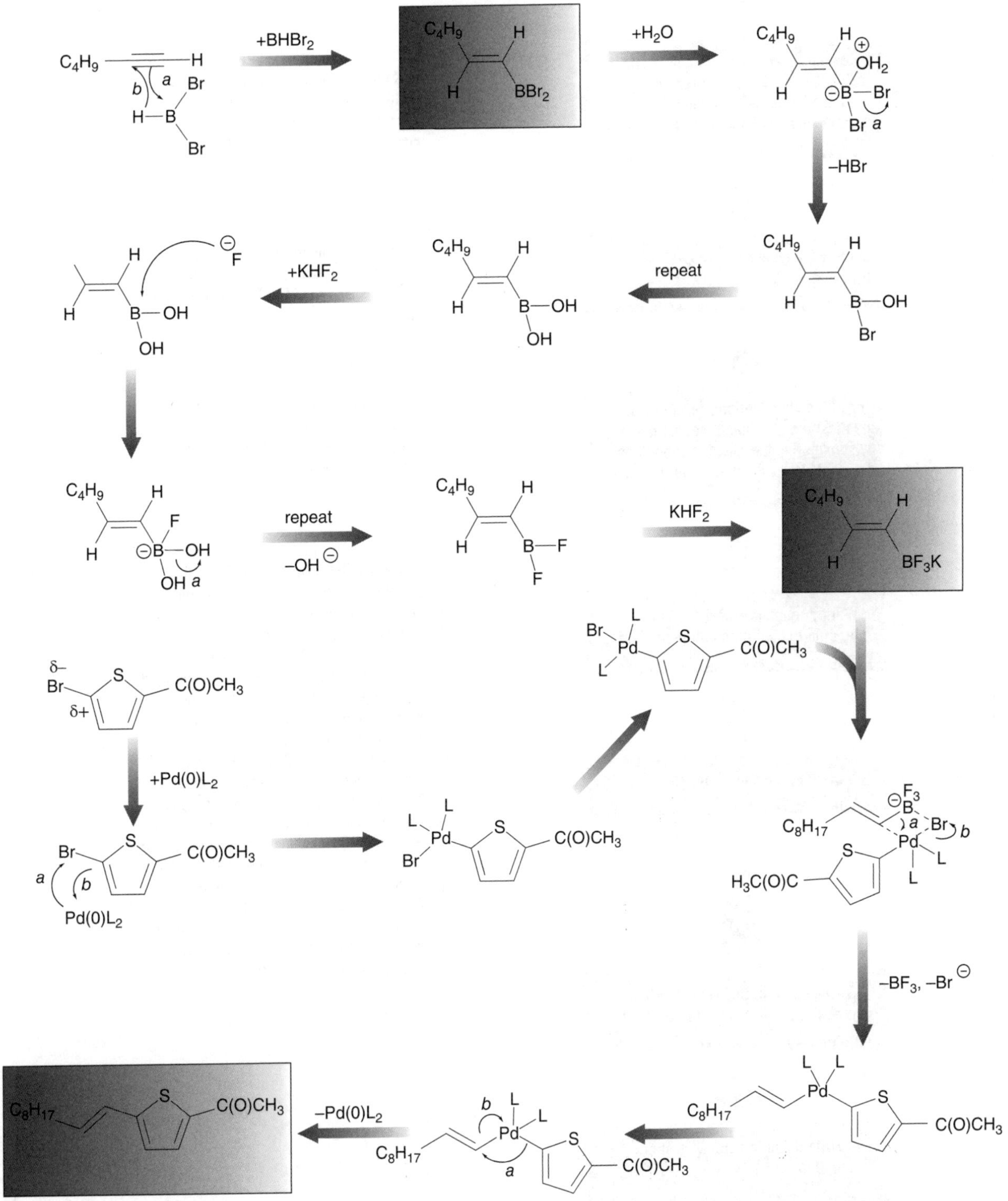

C$_8$H$_{17}$ —$\overset{\delta-}{\equiv}$— H 2 1

$\overset{\delta-}{H}$—B$\overset{\delta+}{}$ Br / Br

$\overset{\delta-}{Br}$ $\overset{}{}$ S 4 C(O)CH$_3$ $\overset{\delta+}{}$ 3

·· Pd(0)

C$_8$H$_{17}$ 1 S 4 C(O)CH$_3$ 2 3

C$_4$H$_9$ —≡— H b a Br / H—B / Br **+BHBr$_2$** → C$_4$H$_9$, H, H, BBr$_2$ **+H$_2$O** → C$_4$H$_9$, H, $\oplus$OH$_2$, B$^{\ominus}$, Br, Br a

−HBr

C$_4$H$_9$, H, H, B, OH, Br

repeat → C$_4$H$_9$, H, H, B—OH, OH

+KHF$_2$ ← $^{\ominus}$F → H, H, H, B—OH, OH

C$_4$H$_9$, H, H, F, B$^{\ominus}$, OH, OH a **repeat −OH$^{\ominus}$** → C$_4$H$_9$, H, H, B—F, F **KHF$_2$** → C$_4$H$_9$, H, H, BF$_3$K

$\overset{\delta-}{Br}$ S C(O)CH$_3$ $\overset{\delta+}{}$ **+Pd(0)L$_2$** ↓ Br, S, C(O)CH$_3$, a b, Pd(0)L$_2$ → L, Pd, Br, S, C(O)CH$_3$ → Br, Pd, L, L, S, C(O)CH$_3$

C$_8$H$_{17}$, $^{\ominus}$BF$_3$, Br a b, Pd, L, L, H$_3$C(O)C, S

−BF$_3$, −Br$^{\ominus}$

C$_8$H$_{17}$, L, L, Pd, S, C(O)CH$_3$

C$_8$H$_{17}$, b, L, L, Pd, a, S, C(O)CH$_3$ **−Pd(0)L$_2$** → C$_8$H$_{17}$, S, C(O)CH$_3$

Now try question 7.18

7.13

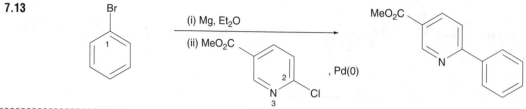

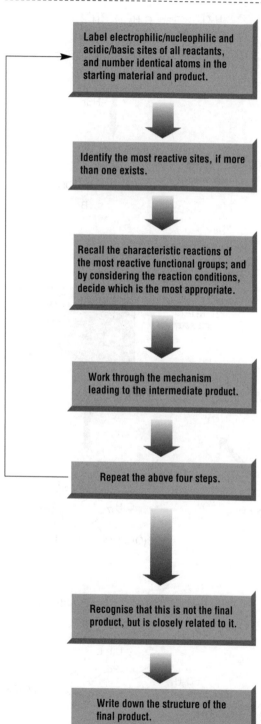

Label electrophilic/nucleophilic and acidic/basic sites of all reactants, and number identical atoms in the starting material and product.	Mg metal is able to oxidatively insert into C–X bonds. Pd(0) is readily oxidised to Pd(II), and a polarised C–Cl bond is capable of accepting the released electrons, to form a Pd–C bond.
Identify the most reactive sites, if more than one exists.	In this case, only reactive sites are the C–Cl and C–Br bonds in each of the starting materials, and the Mg(0) and Pd(0) metal species.
Recall the characteristic reactions of the most reactive functional groups; and by considering the reaction conditions, decide which is the most appropriate.	Palladium(0) mediates the coupling of organomagnesiums (Grignard reagents) and organochlorides in the Kumada (see Negishi) reaction.
Work through the mechanism leading to the intermediate product.	Organohalides (usually with Csp^2–X bonds) and Pd(0) readily participate in lig and coupling processes. Pd(0) is readily oxidised to Pd(II), with simultaneous oxidative addition into the C–Cl bond, to give a *cis*–organopalladium complex, which rapidly equilibrates to the *trans*-system. The bromoalkene reacts with magnesium metal by oxidative addition, to give an organomagnesium (Grignard) species, which is activated to ligand exchange. The *trans*-palladium complex undergoes transmetalation (metal–metal exchange) with the organomagnesium species to give another palladium complex, with the two principal ligands in a *trans*-arrangement. Equilibration to the *cis*-system permits reductive elimination, giving the product by ligand coupling, and releases $Pd(0)L_2$ which then catalyses the next reaction.
Repeat the above four steps.	
Recognise that this is not the final product, but is closely related to it.	Not needed here.
Write down the structure of the final product.	

Now try question 7.19

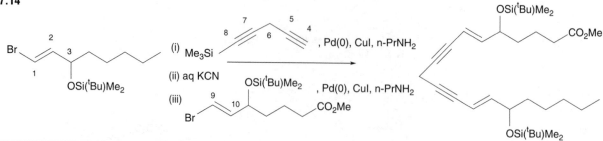

Label electrophilic/nucleophilic and acidic/basic sites of all reactants, and number identical atoms in the starting material and product.	Pd(0) is readily oxidised to Pd(II), and a polarised C–Br bond is capable of accepting the released electrons, to form a Pd–C bond. The acetylene function is acidic, and readily deprotonated.
Identify the most reactive sites, if more than one exists.	In this case, the reactive sites are the C–Br bond, the alkyne and the Pd(0) metal species.
Recall the characteristic reactions of the most reactive functional groups; and by considering the reaction conditions, decide which is the most appropriate.	Palladium(0) mediates the coupling of acetylenes and organoiodides in the Sonagashira reaction.
Work through the mechanism leading to the intermediate product.	*Organohalides (usually with Csp^2–X bonds) and Pd(0) readily participate in ligand coupling processes. Pd(0) is readily oxidised to Pd(II), with simultaneous oxidative addition into the C–Br bond, to give a *cis*–organopalladium complex, which rapidly equilibrates to the *trans*-system. Triethylamine base deprotonates the alkyne, to give an acetylide anion, which reacts with the copper iodide to give a copper acetylide, which is activated to ligand exchange. The *trans*-palladium complex undergoes transmetalation (metal–metal exchange) with the copper acetylide to give another palladium complex, with the two principal ligands in a *trans*-arrangement. Equilibration to the *cis*-system permits reductive elimination, giving the product by ligand coupling, and releases Pd(0)L$_2$ which then catalyses the next reaction.
Repeat the above four steps.	
Recognise that this is not the final product, but is closely related to it.	* Cyanide is an excellent nucleophile that attacks silicon and releases acetylide anion; this reacts with CuI to give the corresponding copper acetylide.
Write down the structure of the final product.	* The Songashira coupling process is then repeated as outlined in the first stage.

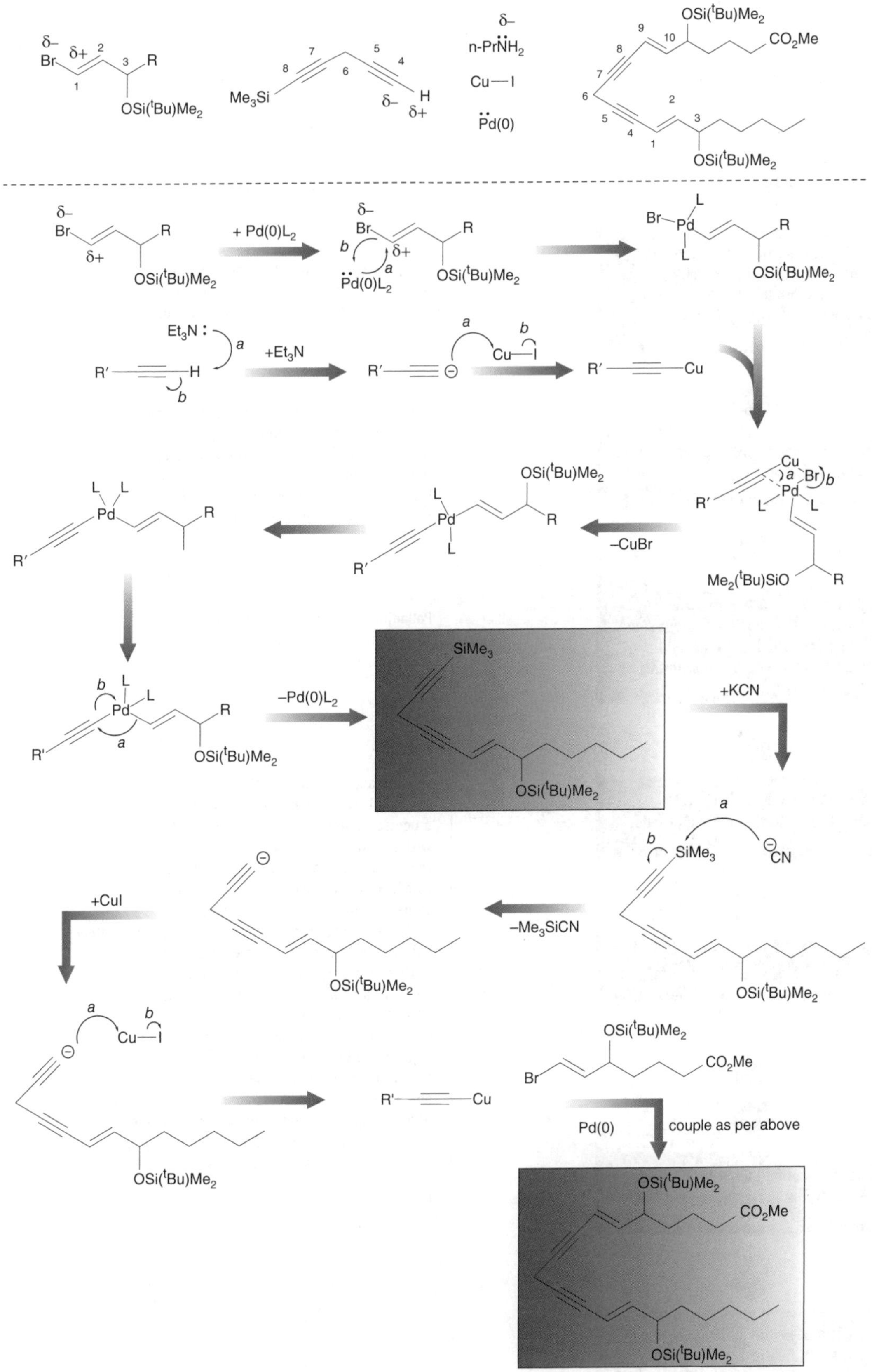

δ−
δ+ 2
Br 3 R
1
OSi(^{t}Bu)Me$_2$

7 5
Me$_3$Si 8 6 4
δ− H
δ+

δ−
n-PrNH$_2$
Cu—I
Pd(0)

OSi(^{t}Bu)Me$_2$
9 10
8 CO$_2$Me
7
6
5 2
4 1 3
OSi(^{t}Bu)Me$_2$

δ−
Br R
δ+
OSi(^{t}Bu)Me$_2$

+ Pd(0)L$_2$

δ−
Br R
b δ+ a
Pd(0)L$_2$ OSi(^{t}Bu)Me$_2$

L
Br Pd R
L
OSi(^{t}Bu)Me$_2$

Et$_3$N: a
R' H +Et$_3$N
b

a b
R' ⊖ Cu—I

R' Cu

Cu Br)b
a
R' Pd
L
Me$_2$(^{t}Bu)SiO R

L L OSi(^{t}Bu)Me$_2$
Pd R L
R Pd R −CuBr
R' OSi(^{t}Bu)Me$_2$ R' L

L L
b
R' Pd
a L R
OSi(^{t}Bu)Me$_2$

−Pd(0)L$_2$

SiMe$_3$
OSi(^{t}Bu)Me$_2$

+KCN

a
b
SiMe$_3$ ⊖ CN

+CuI
⊖

−Me$_3$SiCN

OSi(^{t}Bu)Me$_2$

SiMe$_3$
OSi(^{t}Bu)Me$_2$

a b
⊖ Cu—I

R'—Cu

OSi(^{t}Bu)Me$_2$

OSi(^{t}Bu)Me$_2$
Br CO$_2$Me

Pd(0) couple as per above

OSi(^{t}Bu)Me$_2$
CO$_2$Me

OSi(^{t}Bu)Me$_2$

Now try question 7.20

221

7.15

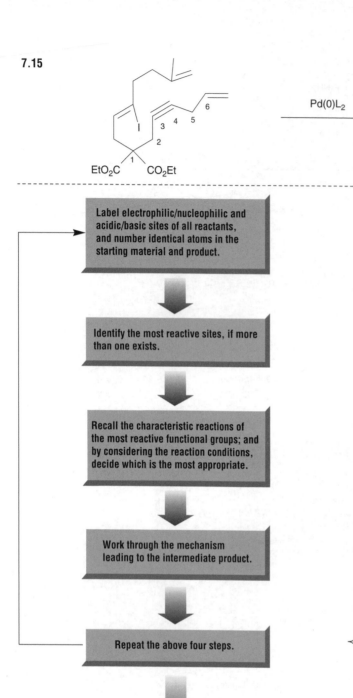

Pd(0)L$_2$

Label electrophilic/nucleophilic and acidic/basic sites of all reactants, and number identical atoms in the starting material and product.

Pd(0) is readily oxidised to Pd(II), and a polarised C–I bond is capable of accepting the released electrons, to form a Pd–C bond.

⬇

Identify the most reactive sites, if more than one exists.

In this case, reactive sites are the C-I bond, the alkyne and the Pd(0) metal species. Adjacent alkene groups are not able to react in the first instance because this would not lead to the formation of the favoured 6-membered ring, but larger ones.

⬇

Recall the characteristic reactions of the most reactive functional groups; and by considering the reaction conditions, decide which is the most appropriate.

Palladium(0) mediates the coupling of alkenes and organoiodides in the Heck reaction.

⬇

Work through the mechanism leading to the intermediate product.

⬇

Repeat the above four steps.

Organohalides (usually with Csp^2–X bonds) and Pd(0) readily participate in ligand coupling processes. Pd(0) is readily oxidised to Pd(II), with simultaneous oxidative addition into the C–I bond, to give *cis*-organopalladium complex, which rapidly equilibrates to the *trans*-system. the *trans*–palladium complex undergoes addition to the most immediately adjacent carbon-carbon triple bond, in a process called carbopalladatiion, to give a new organopalladium species. The resulting organopalladium species is now able to undergo carbopalladation again, now with the adjacent carbon–carbon double bond to give a new organopalladium species, which in turn is able to repeat this process again. β–Elimination of a palladium hydride gives the product, and releases Pd(0)HIL$_2$ which eliminates HI to give which then catalyses the next reaction.

⬇

Recognise that this is not the final product, but is closely related to it.

Not needed here.

⬇

Write down the structure of the final product.

Now try questions 7.21

Index

How to Solve Organic Reaction Mechanisms: A Stepwise Approach, First Edition. Mark G. Moloney.
© 2015 John Wiley & Sons, Ltd. Published 2015 by John Wiley & Sons, Ltd. Companion website: www.wiley.com/go/moloney/mechanisms